Molecular Biology Intelligence Unit

AMINOPEPTIDASES

Allen Taylor, Ph.D.

Tufts University
Boston, Massachusetts, U.S.A.

Springer-Verlag Berlin Heidelberg GmbH

MOLECULAR BIOLOGY INTELLIGENCE UNIT
AMINOPEPTIDASES

R.G. LANDES COMPANY
Austin, Texas, U.S.A.

International Copyright © 1996 Springer-Verlag Berlin Heidelberg

Softcover reprint of the hardcover 1st edition 1996

 Springer

While the authors, editors and publisher believe that drug selection and dosage and the specifications and usage of equipment and devices, as set forth in this book, are in accord with current recommendations and practice at the time of publication, they make no warranty, expressed or implied, with respect to material described in this book. In view of the ongoing research, equipment development, changes in governmental regulations and the rapid accumulation of information relating to the biomedical sciences, the reader is urged to carefully review and evaluate the information provided herein.

Library of Congress Cataloging-in-Publication Data

Taylor, Allen, 1946-
  Aminopeptidases / Allen Taylor.
    p.cm.— (Molecular biology intelligence unit)
  Includes bibliographical references and index.
  ISBN 978-3-662-21605-7 ISBN 978-3-662-21603-3 (eBook)
  DOI 10.1007/978-3-662-21603-3

  1. Aminopeptidases. I. Title. II. Series.
QQP609.A44T38 1996   96-47143
572'.7548--dc21   CIP

PUBLISHER'S NOTE

R.G. Landes Company publishes six book series: *Medical Intelligence Unit, Molecular Biology Intelligence Unit, Neuroscience Intelligence Unit, Tissue Engineering Intelligence Unit, Biotechnology Intelligence Unit* and *Environmental Intelligence Unit.* The authors of our books are acknowledged leaders in their fields and the topics are unique. Almost without exception, no other similar books exist on these topics.

Our goal is to publish books in important and rapidly changing areas of bioscience and environment for sophisticated researchers and clinicians. To achieve this goal, we have accelerated our publishing program to conform to the fast pace in which information grows in bioscience. Most of our books are published within 90 to 120 days of receipt of the manuscript. We would like to thank our readers for their continuing interest and welcome any comments or suggestions they may have for future books.

Shyamali Ghosh
Publications Director
R.G. Landes Company

DEDICATION

To my mother who made every day of childhood wonderful, and to my father whose toil, idealism, devotion, and song nurtured us and whose memory is carved forever in our hearts. To friends, family and colleagues, without whom I would not have had the opportunity to do this book.

CONTENTS

1. **Aminopeptidases, Occurrence, Regulation and Nomenclature ... 1**
   Allen Taylor
   Introduction 1
   Nomenclature 2
   Physiological Function and Uses 5
   Biochemical and Industrial Uses for Aminopeptidases 8
   Regulation of Expression 10
   Summary and Conclusions 11

2. **Structure and Function of Bovine Lens Aminopeptidase**
   and Comparison with Homologous Aminopeptidases 21
   Allen Taylor, David Sanford and Thomas Nowell
   Introduction 21
   Composition and Structure 22
   Inhibitor Binding Site 29
   Metal Ions in Structure and Catalysis 45
   Mechanism of Action 49
   Homologies 52
   Expression 56
   Remaining Mysteries and Conclusions 59

3. **Metallobiochemistry of Aminopeptidases 69**
   Harold E. Van Wart
   Introduction 69
   Classification by Metal Center 69
   The M1 Family of One-Zinc APs 74
   The Two-Zinc Class of APs 77
   The M24 Family of Methionyl APs 85
   Conclusions 87

4. **Methionine Aminopeptidase: Structure and Function 91**
   Ralph A. Bradshaw and Stuart M. Arfin
   Introduction 91
   Cotranslational Processing of Eukaryotic Proteins 92
   Prokaryotic and Yeast MetAP 93
   Higher Eukaryotic MetAPs 97
   Relationship of Human MetAP to Other Proteins 100
   Two Families of MetAPs 101
   Summary and Future Prospective 103

5. **Genetic and Biochemical Analysis of Yeast Aminopeptidases 107**
   Yie-Hwa Chang
   Cytosolic Nonspecific Aminopeptidases 107
   Yeast Vacuolar Aminopeptidases 112
   Methionine Aminopeptidases (MetAPs) 115
   Concluding Remarks 122

6. Alternate Functions for Aminopeptidases:
   Hydrolysis of Leukotriene A$_4$ 129
   F. A. Fitzpatrick and Lars Orning
   Lipid Mediator Biosynthesis by an Aminopeptidase Enzyme 129
   Leukotrienes: Convergence of Lipid and Peptide Enzymology 131
   Enzymology of LTB$_4$ Formation 131
   cDNA Cloning of LTA$_4$ Hydrolase 134
   Recognizing an Aminopeptidase 135
   The Zinc Content and Catalytic Traits of LTA$_4$ Hydrolase 136
   Pharmacological Inhibitors of Aminopeptidase/ LTA$_4$ Hydrolase 138
   Peptide Substrates for Aminopeptidase/LTA$_4$ Hydrolase 142
   Constitutive Regulation of Aminopeptidase E.C. 3.3.2.6
      by LTA$_4$ Hydrolysis 147
   Summary: Past and Future 149

7. Physiological Roles of Ectoenzymes Indicated
   by the Use of Aminopeptidase Inhibitors 155
   Takaaki Aoyagi
   Introduction 155
   Membrane Aminopeptidases 156
   Aminopeptidases in Mammalian Serum and Organs 157
   Aminopeptidase Inhibitors 157
   Role of Ectoenzymes and their Inhibitors in Pathologic Conditions .. 164
   Pharmacologic Applications of Enzyme Inhibitors 166
   Summary 167

8. Plant Aminopeptidases: Occurrence, Function
   and Characterization 173
   Linda L. Walling and Yong-Qiang Gu
   Introduction 173
   Classification of Plant Aminopeptidases 175
   Neutral Aminopeptidases 181
   Alkaline Aminopeptidases 184
   Plant Leucine Aminopeptidases 185
   Plant Cathepsin H Homologs 190
   Role of Plant Aminopeptidases in Plant Development 191
   Seed Germination 191
   Seed and Fruit Maturation 194
   Seedling Growth and Development 195
   Flower Development 200
   Role of Aminopeptidases in Host-Pathogen Interactions 201
   Plant Aminopeptidases in Response to Abiotic Stress 205
   Future Directions 206

Index 219

EDITOR

Allen Taylor, Ph.D.
Jean Mayer USDA Human Nutrition Research Center
on Aging at Tufts University
Laboratory for Nutrition and Vision Research
Boston, Massachusetts, U.S.A.
Chapters 1, 2

CONTRIBUTORS

Takaaki Aoyagi, Ph.D.
Professor of Hygenic Chemistry
Showa College of Pharmaceutical
  Sciences
Machida-shi, Tokyo, Japan
and Director of Enzymology (Guest)
Institute of Microbial Chemistry
Shinagawa-ku, Tokyo, Japan
Chapter 7

Stuart M. Arfin
Department of Biological Chemistry
College of Medicine
University of California
Irvine, California, U.S.A.
Chapter 4

Ralph A. Bradshaw
Department of Biological Chemistry
College of Medicine
University of California
Irvine, California, U.S.A.
Chapter 4

Yie-Hwa Chang
Edward A. Doisy Department of
  Biochemistry & Molecular Biology
St. Louis University School
  of Medicine
St. Louis, Missouri, U.S.A.
Chapter 5

F. A. Fitzpatrick
Department of Pharmacology
University of Colorado Health
  Science Center
Denver, Colorado, U.S.A.
Chapter 6

Yong-Qiang Gu
Department of Botany and Plant
  Sciences
University of California
Riverside, California, U.S.A.
Chapter 8

Thomas Nowell
Jean Mayer USDA Human Nutrition
  Research Center on Aging
  at Tufts University
Laboratory for Nutrition and Vision
  Research
Boston, Massachusetts, U.S.A.
Chapter 2

Lars Orning
Department of Pharmacology
University of Colorado Health
  Science Center
Denver, Colorado, U.S.A.
Chapter 6

David Sanford
Biochemistry Department
Tufts University
Boston, Massachusetts, U.S.A.
Chapter 2

Harold E. Van Wart
Roche Bioscience
Palo Alto, California, U.S.A.
Chapter 3

Linda L. Walling
Department of Botany and Plant
  Sciences
University of California
Riverside, California, U.S.A.
Chapter 8

PREFACE

Aminopeptidases catalyze the cleavage of amino acids from the amino-terminus of protein or peptide substrates. There are over 1000 recent citations regarding the aminopeptidases. They are widely distributed throughout the animal and plant kingdoms. They are found in many subcellular organelles, in cytoplasm, and as membrane components. Mutation of several aminopeptidases is lethal to cells, clearly indicating essential cellular functions for these enzymes. Other aminopeptidases also perform important functions; however, essentiality has not been established. In addition, they are used for diagnostic tests and in industry. Several aminopeptidase structures have been solved, mechanistic information has progressed significantly, and many functions, including dual functions for aminopeptidases, have been elucidated. A variety of protein and gene sequences have become available for aminopeptidases, and the observed homologies between many aminopeptidases from evolutionarily disparate organisms indicate conservation of structure. Despite the burgeoning interest in these enzymes there is no recent compilation of data regarding the aminopeptidases. This book fills this information void.

In the first chapter, nomenclature, biological uses, occurrence in the natural kingdoms and regulation of expression of the enzymes are considered. This is followed by an in-depth consideration of structural and mechanistic features, including roles for metal ions, of the best studied aminopeptidase, leucine aminopeptidase from bovine lens. Homologies between this enzyme and other molecules are also discussed. In the third chapter roles for metals in various aminopeptidases are described. In the fourth and fifth chapters structure and function, and molecular genetics of yeast aminopeptidases are described. In the sixth chapter alternate functions for aminopeptidases such as hydrolysis of leukotriene A_4 are examined. Physiological roles of ectoenzymes as indicated by the pharmacological use of aminopeptidase inhibitors is the topic of chapter 7. In the last chapter, a summary of functions and structural data for plant aminopeptidases and comparisons to other aminopeptidases is presented. As the first summary of information regarding this class of enzymes, the book attempts to be inclusive and focused. I hope that the book sets the stage for future work. Space limitations prevented this text from being even more exhaustive in coverage of the literature, and we apologize to authors whose good work could not be given more attention.

CHAPTER 1

AMINOPEPTIDASES, OCCURRENCE, REGULATION AND NOMENCLATURE

Allen Taylor

INTRODUCTION

Aminopeptidases (AP) catalyze the hydrolysis of amino acid residues from the amino terminus of peptide substrates. These enzymes generally have broad specificity, occur in several forms, and are widely distributed throughout the plant and animals kingdoms. Over 100 APs have been purified and/or studied, and over 50 genes have been cloned and characterized. Several forms of these enzymes have been found in many tissues or cells, on cell surfaces, and in soluble cytoplasmic or secreted forms in plants and animals[1-7] (see chapters 2-8). In some cells they constitute a substantial proportion of enzyme protein.[4,8,9]

Several physiological functions of APs have been identified. Aminopeptidases are essential for protein maturation,[10] degradation of nonhormonal[11] and hormonal peptides, and, possibly, determination of protein stability.[12] Many disease states are associated with impaired proteolytic function.[13-19] Several APs also catalyze reactions in addition to peptide hydrolysis. Recent data regarding these subjects will be described below and in the accompanying chapters.

Aminopeptidases, edited by Allen Taylor. © 1996 R.G. Landes Company.

NOMENCLATURE

It would appear *de rigueur* that any text regarding a specific group of molecules should begin with a classification of members of that group. However, a unique nomenclature system for APs remains elusive. Instead, classification of AP has been done primarily according to the interest of the investigator. This results in a practical, if labyrinthian, series of nonexclusive names.[1,20-22] The following criteria have been used in various classification schemes.

NUMBER OF AMINO ACIDS CLEAVED FROM THE N-TERMINUS OF SUBSTRATES

Enzymes which sequentially remove the NH_2-terminal amino acid from protein and peptide substrates are called aminopeptidases. Aminodipeptidases, or diaminopeptidases, remove intact NH_2-terminal dipeptides. Aminotripeptidases catalyze the hydrolysis of NH_2-terminal tripeptides.[1,22] These are given the numbers E.C. 3.4.11-3.4.13.

WITH RESPECT TO THE RELATIVE EFFICIENCY WITH WHICH NH_2-TERMINAL RESIDUES ARE REMOVED FROM PEPTIDES OR PEPTIDE ANALOGS

Leucine aminopeptidases (LAP) remove most effectively Leu and other hydrophobic residues from peptide substrate analogs, although bonds to many other residues are cleaved. Indeed, the complete degradation of a protein by LAP has been described.[8] Arginine-, methionine-, aspartic-, alanine-, glutamic-, proline- and cystinyl-APs have also been described. In most instances, the kinetic tests used for identification were done with amino analogs (i.e., amino acyl-p-nitroanilides or β-naphthylamides) as substrates. However, these analogs are rarely hydrolyzed at rates comparable to rates of hydrolysis of peptides which bear the same amino terminal residue.[23] In some cases the peptidase name includes information regarding the substrate analog used, i.e., leucyl-β-naphthyl amidase, although this is a physiologically irrelevant substrate. In addition, metal ion content in many enzymes has rarely been held constant. Each of these influences relative activities, making classification according to the residue cleaved a nettlesome chore[23,24] (see Table 2.4 in chapter 2). For example, lens LAP in the $Zn^{2+}Mg^{2+}$ form (see metal content below) shows a specific activity for a typical physiological substrate, LeuGlyGly, of 804 μmol/min/mg and a K_m of 1 mM. In contrast, for the widely used substrate leucyl-p-nitroanilide (LpNA), the specific activity is 3.9 µmol/min/mg; yet K_m is similar (approximately 3 mM) to that for the peptide. In comparison, AlaGlyGly is also rapidly hydrolyzed (specific activity 604 µmol/min/mg), although the K_m is 50 times as great.[23] Analogous information regarding porcine kidney LAP was obtained by Van Wart and coworkers.[25] The difficulty in utilizing substrates for characterization is corroborated by Patterson's[26] observation of different recovery times for bestatin-inhibited aminodipeptidase utilizing different substrates.

In most cases the αNH$_2$-residue must be of the L-configuration. However, Asano et al purified from *Ochrobactrum anthropi* SCRC C1-38 a dimeric (59 kDa/subunit) aminopeptidase specific for D amino acids.[27] This enzyme is not inhibited by metal-chelating agents, but is inhibited by sulfhydryl reagents.

LOCATION

Some peptidases are secreted,[28] but most are not. There are cytosolic and microsomal enzymes, and integral-membrane-bound or membrane-associated enzymes.[20,21,29] Membrane anchors include glycosylphosphatidylinositol.[30] Other APs are found in organelles, such as lysosomes, nuclei, and mitochondria. A problem with current nomenclature which uses trivial names is that in some instances, different enzymes share the same name. It is not uncommon for structurally unrelated enzymes to have the same name and yet to be found in different locations within one organism, i.e., membrane-bound and methionine aminopeptidases have both been referred to as AP-M. Thus, even locations of the enzymes could be more clearly described in the name. In addition, aspects of the enzyme-membrane interaction that alter activity remain to be elucidated.

INHIBITOR SUSCEPTIBILITY

Aminopeptidases may be inhibitable by bestatin, amastatin, or other transition-state analogs (see chapter 2 for structures and inhibition constants of these inhibitors). The most commonly used aminopeptidase inhibitor, bestatin, is an analog of the dipeptide substrate PheLeu (see chapter 2). It contains an α-carbon bearing a hydroxyl group which mimics the putative transition state in peptide hydrolysis. Incorporation of this functional group enhances binding of the inhibitor $\approx 10^5$ (see chapter 2). It is suggested that the tighter binding of polypeptide inhibitors such as amastatin, in preference to dipeptide analogs, can be used to distinguish membrane-bound (i.e., AP-M) from cytosolic enzymes.[31] Metal ion chelators and functional group specific reagents are also effective inhibitors of various APs.

METAL ION CONTENT

There are both metal ion-containing and metal-free APs. For those APs which incorporate metal ions, the ions may or may not be directly involved in catalysis[20,32] (see chapter 2). Metal ions which are involved in AP structure and function include Co^{2+} and Zn^{2+}. Residues which bind the metal to the enzyme may also be used to describe APs[33-35] (see chapters 2 and 3).

CONDITIONS OF MAXIMAL ACTIVITY

The pH at which maximal activity is observed has often been used to characterize APs. There are acidic, basic, and "neutral" peptidases. Some APs are chloride, or other ion, activated.

SIZE

Monomeric, polymeric, low-, and high-mass peptidases have been documented. Most,[20,21] but not all[36-38] polymeric APs are homopolymers, i.e., polymers of a given protomer. This is contrasted with high mass endopeptidases (proteasomes) which are comprised of a myriad of different protomers.

THERMOSTABILITY

Aminopeptidases which can withstand elevated temperatures have been identified. For example, *Aeromonas* has a "thermostable" AP[39] (see Table 2.1 in chapter 2 and descriptions of thermosensitive APE2 in chapter 5). However, most other APs do not withstand prolonged exposure to elevated temperatures and cannot be classified as thermostable.

NUMBER OF FUNCTIONS

While for most aminopeptidases only one function has been elucidated, there are now several APs for which two functions have been demonstrated (see chapters 5, 6, 8, and below).

Obviously, each of these nomenclature or classification systems serves to identify the protease with respect to a topic of interest, and they are not mutually exclusive. Given the labyrinth of classifications, it is not surprising that in recent years several enzymes, which were previously thought to be distinct, were shown to be identical (see chapter 2). For example, porcine kidney LAP (which was described as not able to hydrolyze prolyl bonds) and porcine intestinal prolyl AP are indistinguishable.[40] The same pertains to rat kidney and brain prolyl AP and porcine kidney LAP.[23,24,41,42] The *E. coli xerB* gene product, *E. coli* AP-I, now called AP-A, and *S. typhimurium* aminopeptidase appear to be the same,[5] as are APs N and M (E.C. 3.4.11.2).[1] Perhaps it would be advantageous to assay an AP with many peptide substrates prior to choosing a name.

PHYSIOLOGICAL FUNCTION AND USES

Some aminopeptidase activities are essential. Others, while not essential, affect cells in profound ways. In all living cells, protein synthesis is initiated at an AUG codon. This initiation AUG codon specifies methionine in the cytosol of eukaryotes, or formyl-methionine in prokaryotes, mitochondria, and chloroplasts. Protein maturational events, including NH_2-terminal modifications of nascent peptides, are by far the most common processing events, occurring on nearly all proteins.[10] In this process where formyl-methionine is used to initiate protein synthesis, the formyl group is usually removed cotranslationally by a deformylase, leaving methionine bearing a free NH_2 group.[10,12,43-47] In both eukaryotes and prokaryotes, the NH_2-terminal methionine may be removed by a methionine aminopeptidase(s) (MAP).[10,12,47-49] Removal of the NH_2-terminal methionine is required in order to reveal functionally important NH_2-terminal residues and/or to allow NH_2-terminal modification (such as myristoylation) which is required for physiologic function.

Essentiality of certain APs is indicated, since deletion of those enzymes is lethal. For example, deletion of methionine AP (*pepM* gene product) is lethal in *Salmonella typhimurium*.[50] Consistent with essentiality of some methionine APs is the observation that *pepM* mutants of *E. coli* or *S. typhimurium* could not be obtained.[49,50] Since, except in disease states or upon aging, protein fragments rarely accumulate,[13,14] a "housekeeping" role for APs is indicated in continuous protein turnover and/or regulation of protein levels, selective elimination of obsolete or defective proteins, supply of amino acids and energy during starvation and/or differentiation, and degradation of transported exogenous peptides to amino acids for nutrition. Indeed, experiments using the relatively tight-binding inhibitor, bestatin, indicate that APs are required in mammalian tissues and cells for terminal stages of intra- and extracellular protein degradation[11] and for EGF-induced cell-cycle control.[51] Proteolytic capabilities also allow cells to adapt to changing environmental conditions. This is confirmed by the time-related reduction in viability of some strains mutated for several AP coding genes.[52] In *E. coli*, AP-A (pepA, the *xerB* gene product) is absolutely required for ColE1 stabilization of unstable plasmid multimers, which occurs via site-specific recombination (at the cer locus) into monomer form.[5] This is also identical with the carP gene which is involved in pyrimidine-specific regulation of the upstream P1 promoter of *E. coli*.[53]

In contrast with the essentiality of the APs noted above, in *Saccharomyces* deletion of the *MAP1* gene is associated with retarded growth but is not lethal.[54] This suggests that alternative NH_2-terminal processing pathways exist for cleaving methionine from nascent polypeptide chains in eucaryotic cells.[7,55] Redundancy of AP activities in procaryotic and eucaryotic cells may provide yet another, albeit teleological, support for the important cellular functions for APs.[7,55]

*BLH*1 codes for a yeast thiol aminopeptidase which is homologous to mammalian bleomycin hydrolase.[56] Whereas deletion of the *BLH*1 gene is not lethal under normal growth conditions, blh1 mutants show hypersensitivity to bleomycin. This indicates that bleomycin hydrolase is able to inactivate bleomycin in vivo and to protect cells from bleomycin-induced toxicity.[56]

To the extent to which physiological functions which are affected by bestatin involve aminopeptidases, it would appear that APs are involved in the delayed-type hypersensitivity,[16] murine tumor growth rate and enhanced antitumor activity of antibiotics,[57] DNA metabolism in spleen and thymus T-cells,[58] and stimulated polysome assembly.[59] Roles for AP in antibiotic activation and transport have been documented.[29]

Leukotriene hydrolase activity of an aminopeptidase suggests roles for APs in inflammation[60-62] (see chapter 6). Also suggestive of relations between AP activity and inflammation are observations that AP-P (aminoacylprolyl-peptide hydrolase E.C. 3.4.11.9) may be involved in hydrolyzing bradykinin[63,64] (see Table 2.1, chapter 2).

Recent data indicate that bestatin-inhibitable aminopeptidases are involved in the conversion of procollagenase to collagenase.[65] In related research it was demonstrated that AP-N/CD13 plays a role in degradation and invasion of extracellular matrix since A375M melanoma cells transfected with full-length cDNA of AP-N/CD13 show enhanced degradation of type IV collagen.[66] This association between AP activity and tissue invasion is corroborated by data which show that monoclonal antibodies to AP-N/CD13 inhibit invasion of metastatic renal cells into Matrigel-coated filters and degradation of type IV collagen, and that this activity is also inhibitable by bestatin.[67] Aminopeptidase N from pig contains sequences which can act as cellular, virus-binding receptors.[68]

Aminopeptidases also participate in the metabolism of secreted regulatory molecules including hormones and neurotransmitters.[6,28,33,42,69,70] This includes partial degradation of enkephalin by cerebral pericytic AP-N at the blood-brain interface[71] and by murine macrophages. Another function of APs appears to be in modulation of cell-cell interactions (for review see reference 4 and references cited within).

In the blood clotting cascade, aminopeptidase A appears to liberate angiotensin from angiotensin II.[72] Aminopeptidase N participates in inactivation of angiotensin III. This information has been exploited for the design of drugs to regulate rates of blood clotting.

Evidence for dual functions or activities of aminopeptidases has been mounting. In some cases, mutational analysis has been cleverly exploited to separate functions in the APs. For example, mutation of the *PEP A* gene which inactivates the hydrolytic activity does not eliminate the role of AP-A in the recombination process.[73] Second functions for APs, which are separable from AP activity, are also reported for leukotriene A_4 hydrolase[62] (also see chapter 6) and *O. anthropi* D-aminopeptidase (see chapter 2). Dual function may be implied by structural studies of methionine AP from *Saccharomyces cerevisiae*. Whereas a form of MAP missing the Zn^{2+} fingers is as active as the wild type, the truncated form is significantly less active in rescuing the slow growth phenotype of the *map* mutant than wild type MAP[15] (also see chapters 2, 4 and 5). Thus, initiation factor-associated proteins may include methionine aminopeptidases. Another example of a second function for aminopeptidases is found in aminopeptidase N which serves as a coronavirus receptor.[74,75]

The ubiquitin proteolytic pathway is involved in the degradation of normal, obsolete, and damaged proteins.[76-80] In some cases the N-end amino acid is involved in determining the rate of proteolysis of substrates by this pathway.[80] A central role for APs in defining the stability of proteins can be envisioned for those proteins which are degraded by such an "N-end ubiquitin-dependent degradation pathway" (Fig. 1.1). A corollary is that retention of Met or other stabilizing residues may protect short-lived proteins from premature protein degradation.[12,54]

In contrast with the catabolic roles for aminopeptidases, a biosynthetic or hydrolytic role in peptidoglycan has been suggested.[29] (See section below.) Further advances regarding physiological functions of APs should be possible, since the availability of new fluorogenic substrates for aminopeptidases makes it possible to detect their activity in vivo.[81]

BIOCHEMICAL AND INDUSTRIAL USES FOR AMINOPEPTIDASES

Aminopeptidases have frequently been used to sequentially remove amino-terminal amino acids from proteins, i.e., AP-M was used for studies of structure-biologic function of peptides that bind the thrombin receptor.[82] Pyroglutamate AP has found use in determining sequence and content of pyroglutamate in—and for deprotection prior to sequencing of—proteins.[83,84] In the dairy industry the terminal degradation of peptides derived from casein is accomplished with APs.[29]

Aminopeptidases have also found use in peptide synthesis. Prolyldipeptidyl AP from lactococcus lactis (PepX) was used as a catalyst in the kinetically controlled synthesis of peptide bonds involving proline.[85] Aminopeptidase A was used for reversible protection of the αNH_2 group of amino acids.[86]

An unanticipated observation is that aminopeptidase N binds concanavalin A and promotes cholesterol crystallization.[87]

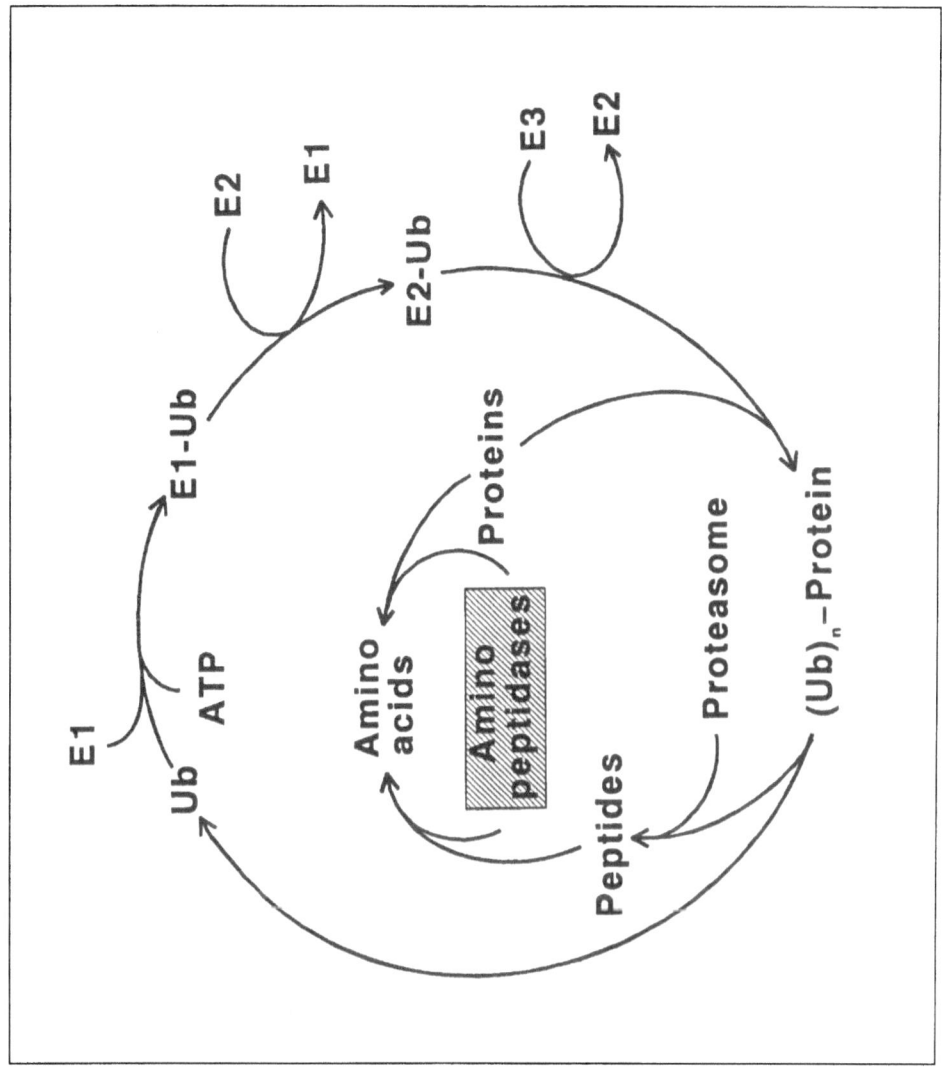

Fig. 1.1. Schematic of possible role of aminopeptidases in determining rates of degradation of proteins. According to the "N-end rule," the rate at which a protein is degraded is determined (at least in part) by its amino-terminal amino acid residue.[12,80] Since aminopeptidases remove amino-terminal amino acids, thus revealing new amino-terminal amino acid residues, the aminopeptidases may play significant roles in determining rates of protein degradation. Thus, the aminopeptidases are shown in a central position in this schematic of the ubiquitin-dependent proteolytic pathway. E1, E2 and E3 are enzymes which activate and sequentially transfer ubiquitin to protein substrates. See references 76-80 for further details regarding the ubiquitin-dependent proteolytic pathway.

REGULATION OF EXPRESSION

Several experiments are consistent with regulation of aminopeptidase expression at the transcriptional level. There is a marked enhancement of LAP activity[88] and LAP mRNA[34] upon removal of serum from culture media in which in vitro aged and/or transforming lens epithelial cells were grown. Concerted induction of LAP mRNA and LAP protein by interferon gamma was also noted in human ACHN renal carcinoma, A549 lung carcinoma, HS153 fibroblasts, and A375 melanoma.[89] Induction of LAP mRNA is a secondary response to interferon, blocked by inhibition of protein synthesis.

Membrane-bound AP-N/CD13 is one of several enzymes expressed in a lineage-restricted fashion by subsets of normal and malignant cells. Genetic elements involved in expression of both transcripts for this protease have been identified, and it appears that physically distinct promoters have evolved to regulate expression of this enzyme in different tissues.[90] Keller recently demonstrated that a member of the mammalian zinc-dependent membrane APs is vesicle protein 165, the cellular distribution of which is at least in part insulin regulated.[91] Receptor function for aminopeptidase N has also been confirmed by molecular genetic techniques in *M sexta*.[92]

The yeast *yscI* gene product, APE1, is the vacuolar glycoprotein AP,[55] which appears to facilitate amino acid uptake by hydrolyzing peptides prior to absorption. APE1 synthesis is subject to carbon catabolite levels, i.e., APE1 is repressed in media containing more than 1% glucose. But as cells reach the stationary phase, the increase in APE1 activity may indicate the release from carbon catabolite repression. Other examples of induction by amino acid limitation or catabolite repression,[93-96] as well as during different phases of cell cycle,[97-102] have been noted. Studies regarding the APE1 promoter are in progress.[103] It is curious that the enzyme isolated from stationary cells has four amino acids removed from the N-terminus.[7] Activity is enhanced several fold when ammonia, rather than peptone, is used as the sole source of nitrogen.[104] Expression of yscI is dependent upon levels of yscA and *PEP4* gene product.[105]

Some aminopeptidases are induced by high temperature[37,38,106-110] and during oxygen limitation.[111-115] Thus, deletion analysis was used to show transcriptional regulation for the *E. coli pepN* gene upon anaerobiosis and phosphate starvation.[29,113]

LAP[9] and some plant AP[116] levels are also enhanced during development and growth. In *Lactobacillus* one open reading frame is cotranscribed with the *pep*C gene at the exponential phase of growth, whereas, at the stationary growth phase, transcripts from the *pep*C promoter were below the detection limit, and the ORF2 was expressed by its own promoter.[99]

Many of the bacterial aminopeptidases display promoter consensus sequences characteristic of genes transcribed by an RNA polymerase associated with σ70 (see ref. 29 for a more thorough treatment regarding bacterial APs). Several genes encoding APs are part of operons. Regulation at the post-transcriptional level is reported for *Aeromonas proteolytica* aminopeptidase.[35] The enzyme is synthesized as a 43 kDa precursor. Maintenance of the organism at elevated temperatures results in double cleavages and frees a mature and active 32 kDa enzyme which is active at 70°C. The precursor is active but is inactivated at 70°C.

Regulation can also be affected by many inhibitors which are in abundant supply in many organisms (see chapter 7).

Nonlethal flux of UV (<12 Jm²) to HeLa cell cultures are followed by large increases in cell surface proteolytic activities including aminopeptidase activities.[117] Administration of cycloheximide also resulted in increased AP activity.[117] Indeed, UV and cycloheximide can initiate reactions in HeLa cells which result in ectopeptidase activation of a global nature, with many analogies to—or which may be part of—a genetic stress response.

SUMMARY AND CONCLUSIONS

Aminopeptidases play critical roles in protein maturation, regulation of hormone levels, selective or homeostatic protein turnover, and plasmid stabilization. The enzymes have long been used for diagnosis of various physiological states and disease conditions, and recently they have found industrial use as well.[118] Administration of bestatin, which presumably has its effect by inhibition of APs, has also been used to alter physiological status and disease progress. The availability of further genetic, structural, and kinetic data, as well as substrates which can be used to monitor AP activity in whole cell systems, should aid in elucidating physiological functions and mechanisms of action of these enzymes. A uniform, simple nomenclature system would avoid confusion and duplication of effort. It would appear that activity assays which employ physiologically relevant dipeptides or tripeptides as substrates may be more informative for nomenclature and characterization purposes than assays which utilize synthetic peptide analogs.

ACKNOWLEDGMENTS

This project was funded in part with federal funds from the United States Department of Agriculture under contract number 53-3K06-0-1; National Institutes of Health, grant number EYO8566; The Daniel and Florence Guggenheim Foundation; and the Research Corporation. The author is indebted to Esther Epstein for preparation of the manuscript and Dr. Jessica Jahngen-Hodge for editorial assistance.

REFERENCES
1. McDonald JK, Barrett AJ. A brief history of the study of mammalian exopeptidases. In: McDonald JK, Barrett AJ, eds. Mammalian Proteases. Vol. 2. New York: Academic Press, 1986.
2. Taylor A, Surgenor T, Thomson DKR et al. Comparison of (concentration and amino acid sequence homology between) leucine aminopeptidase in human lens, beef lens, beef kidney, hog lens, and hog kidney. Exp Eye Res 1984; 38:217-29.
3. Ledeme N, Vincent-Fiquet O, Hennon G et al. Human liver l-leucine aminopeptidase: Evidence for 2 forms compared to pig liver enzyme. Biochimie 1983; 65:397-404.
4. Watt VM, Yip CC. Amino acid sequence deduced from a rat kidney cDNA suggests it encodes the Zn^{2+}-peptidase aminopeptidase N. J Biol Chem 1989; 264:5480-7.
5. Stirling VJ, Colloms SD, Collins JF et al. *xerB*, an *Escherichia coli* gene required for plasmid ColE1 site-specific recombination, is identical to *pepA*, encoding aminopeptidase A, a protein with substantial similarity to bovine lens leucine aminopeptidase. EMBO J 1989; 8:1623-7.
6. Ahmad S, Ward PE. Role of aminopeptidase activity of circulating angiotensins. J Pharmacol Exp Ther 1990; 252:643-50.
7. Chang Y-H, Smith JA. Molecular cloning and sequencing of genomic DNA encoding aminopeptidase I from *Saccharomyces cerevisiae*. J Biol Chem 1989; 264:6979-83.
8. Taylor A, Daims MA, Lee J et al. Identification and quantification of inactive leucine aminopeptidase in aged normal and cataractous human lenses and ability of bovine lens LAP to cleave bovine crystallins. Curr Eye Res 1982; 2:47-56.
9. Taylor A, Brown MJ, Daims MA et al. Localization of leucine aminopeptidase in hog lenses using immunofluorescence and activity assays. Invest Ophthalmol Vis Sci 1983; 24:1172-81.
10. Moerschell RP, Hosokawa Y, Tsunawa S et al. The specificities of

yeast methionine aminopeptidase and acetylation of amino-terminal methionine in vivo. J Biol Chem 1990; 265:19638-43.

11. Botbol V, Scornik OA. Measurement of instant rates of protein degradation in the livers of intact mice by the accumulation of bestatin-induced peptides. J Biol Chem 1991; 266:2151-7.

12. Bachmair A, Finley D, Varshavsky A. In vivo half-life of a protein is a function of its amino-terminal residue. Science 1986; 234:179-86.

13. Taylor A, Dorey CK, Jacques PF. Oxidation and aging: impact on vision. In: Williams G, ed. Proc Int Conf on Antioxidants, vol. 8. Princeton, NJ: Princeton Scientific Press, 1993:349-371.

14. Dice JF. Cellular and molecular mechanisms of aging. Phys Rev 1993; 73:149-159.

15. Zuo S, Guo Q, Ling C et al. Evidence that two zinc fingers in the methionine aminopeptidase from *Saccharomyces cerevisiae* are important for normal growth. Mol Gen Genet 1995; 246:247-53.

16. Umezawa H, Ishizuka M, Aoyagi T et al. Enhancement of delayed-type hypersensitivity by bestatin, an inhibitor of aminopeptidase B and leucine aminopeptidase. J Antibiot 1976; 29:857-9.

17. Nishizawa R, Saino T, Takita T et al. Synthesis and structure-activity relationships of bestatin analogues, inhibitors of aminopeptidase B. J Med Chem 1977; 20:510-5.

18. Aoyagi T. Small molecular protease inhibitors and their biological effects. In: Kleinkauf H, Dohren H, eds. Biochemistry of Peptide Antibiotics. Berlin, New York: Walter de Greyter, 1990:311-63.

19. Taylor A, Davies KJA. Protein oxidation and loss of protease activity may lead to cataract formation in the aged lens. Free Radic Biol Med 1987; 3:371-37.

20. Taylor A. Aminopeptidases: toward a mechanism of action. TIBS 1993; 18:167-72.

21. Taylor A. Aminopeptidases: structure and function. FASEB J 1993; 7:290-8.

22. Enzyme Nomenclature. New York: Academic Press, 1978.

23. Taylor A, Tisdell FE, Carpenter FH. Leucine aminopeptidase (bovine lens): synthesis and kinetic properties of ortho, meta, and para substituted leucyl-anilides. Arch Biochem Biophys 1981; 210:90-7.

24. Sanderink G-J, Artur Y, Seist G. Human aminopeptidases: a review of the literature. J Clin Chem Cli Biochem 1988; 26:795-807.

25. Lin W-Y, Van Wart HE. Steady-state kinetics of hydrolysis of dansyl-peptide substrates by leucine aminopeptidase. Biochemistry 1988; 27:5062-8.

26. Patterson EK. Inhibition by bestatin of a mouse ascites tumor dipeptidase. J Biol Chem 1989; 264:8004-11.

27. Asano Y, Nakazawa A, Kato Y et al. Properties of a novel D-stereospecific aminopeptidase from *Ochrobactrum anthropi*. J Biol Chem 1989; 264:14233-39.

28. Squire CR, Talebian M, Menon JG et al. Leucine aminopeptidase-like activity in *Aplysia hemolymph* rapidly degrades biologically active alpha-bag cell peptide fragments. J Biol Chem 1991; 266:22355-63.

29. Gonzales T, Robert-Baudouy J. Bacterial aminopeptidases: properties and functions. FEMS Microbiol Rev 1996, in press.

30. Turner AJ. Structural and immulological studies of GPI-anchored brush border hydrolases. Brazilian J Med Biol Res 1994; 27:389-94.

31. Rich DH, Moon BJ, Harbeson S. Inhibition of aminopeptidases by amastatin and bestatin derivatives. Effect of inhibitor structure on slow-binding processes. J Med Chem 1984; 27:417-22.

32. Strater N, Lipscomb WN. Transition state analogue L-leucine-phosphonic acid bound to bovine lens leucine aminopeptidase: X-ray structure at 1.65 Å resolution in a new crystal form. Biochemistry 1995; 34:9200-10.

33. Malfroy B, Kado-Fong H, Gros C et al. Molecular cloning and amino acid sequence of rat kidney aminopeptidase M: a member of a super family of zinc-metallohydrolases. Biochem Biophys Res 1989; Commun 161:236-41.

34. Wallner BP, Hession C, Tizard R et al. Isolation of bovine kidney leucine aminopeptidase cDNA: Comparison with the lens enzyme and tissue-specific expression of two mRNAs. Biochemistry 1993; 32:9296-301.

35. Rawlings ND, Barrett AJ. Evolutionary families of metallopeptidases. Methods Enzymol 1995; 248:183-228.

36. Shibata K-L, Watanabe T. Purification and characterization of an aminopeptidase from *Mycoplasma salivarium*. J Bacteriol 1987; 169:3409-13.

37. Stoll E, Ericsson LH, Zuber H. The function of the two subunits of thermophilic aminopeptidase I. Proc Natl Acad Sci USA 1973; 70:3781-4.

38. Stoll E, Hermodson MA, Ericsson LH et al. Subunit structure of the thermophilic aminopeptidase I from *Bacillus stearothermophilus*. Biochemistry 1972; 11:4731-5.

39. Guenet C, Lepage P, Harris BA. Isolation of the leucine aminopeptidase gene from *Aeromonas proteolytica*. J Biol Chem 1992; 267:8390-5.

40. Matsushima M, Takahashi T, Ichinose M et al. Purification and characterization of prolyl aminopeptidases from pig intestinal mucosa and human liver. Structural, immunological and enzymatic evidence for identity with leucyl aminopeptidase. Biochem Biophys Res Commun 1991; 178:1459-64.

41. Turzynski A, Mentlein R. Prolyl aminopeptidase from rat brain and kidney. Eur J Biochem 1990; 190:509-15.

42. Gibson AM, Biggins JA, Lauffart B et al. Human brain leucyl aminopeptidase: isolation, characterization, and specificity against

some neuropeptides. Neuropeptides 1991; 19:163-8.

43. Ball LA, Kaesberg P. Cleavage of the N-terminal formylmethionine residue from a bacteriophage coat protein in vitro. J Mol Biol 1973; 79:531-7.

44. Hausman MS, Snyderman R, Mergenhagen SE. Humoral mediators of chemotaxis of mononuclear leukocytes. J Infect Dis 1972; 125:595-602.

45. Takeda M, Webster RE. Protein chain initiation and deformylation in B. subtilis homogenates. Proc Nat Acad Sci USA 1968; 60:1487-94.

46. Tsunasawa S. Amino-terminal processing of nascent proteins: their role and implication on biological function. [Review] Tanpakushitsu Kakusan Koso—Protein, Nucleic Acid, Enzyme. 1995; 40:389-98.

47. Wilcox C, Hu J-S, Olson EN. Acylation of proteins with myristic acid occurs cotranslationally. Science 1987; 238:1275-8.

48. Ben-Basset A, Bauer K, Chang S-Y et al. Processing of the initiation methionine from proteins: properties of the *Escherichia coli* methionine aminopeptidase and its gene structure. J Bacteriol 1987; 169:751-757.

49. Miller CG, Strauch KL, Kukral AM et al. N-terminal methionine-specific peptidase in *Salmonella typhimurium*. Proc Natl Acad Sci USA 1987; 84:2718-22.

50. Miller CG, Kukral AM, Miller JL et al. PepM is an essential gene in salmonella typhimurium. J Bacteriol 1989; 171:5215-17.

51. Takahashi S-I, Ohishi Y, Kato H et al. The effects of bestatin, a microbial aminopeptidase inhibitor, on epidermal growth factor-induced DNA synthesis and cell division in primary cultured hepatocytes of rats. Exp Cell Res 1989; 183:399-412.

52. Reeve CA, Bockman AT, Martin A. Role of protein degradation in the survival of carbon-starved *Escherichia coli* and *Salmonella typhimurium*. J Bacteriol 1984; 157:758-63.

53. Charlier D, Hassanzadeh G, Kholti A et al. carP, involved in pyrimidine regulation of the *Escherichia coli* carbamoylphosphate synthetase operon encodes a sequence-specific DNA-binding protein identical to XerB and PepA, also required for resolution of Co1EI multimers. J Mol Biol 1995; 250:392-406.

54. Chang YH, Teichert U, Smith JA. Molecular cloning, sequencing, deletion, and overexpression of a methionine aminopeptidase gene from *Saccharomyces cerevisiae*. J Biol Chem 1992; 267:8007-11.

55. Cueva R, Garcia-Alvarez N, Suarez-Rendueles P. Yeast vacuolar aminopeptidase yscI isolation and regulation of the *APE1 (LAP4)* structural gene. FEBS Lett. 1989; 259:125-9.

56. Enenkel C, Wolf DH. BLH1 codes for a yeast thiol aminopeptidase, the equivalent of mammalian bleomycin hydrolase. J Biol Chem 1993; 268:7036-43.

57. Ishizuka M, Masuda T, Kanbayashi N et al. Effect of bestatin on

mouse immune system and experimental murine tumors. J Antibiot 1980; 33:642-52.

58. Muller WEG, Zahn RK, Arendes J et al. Activation of DNA metabolism in T-cells by bestatin. Biochem Pharmacol 1979; 28:3131-37.

59. Muller WEG, Zahn RK, Maidhof A et al. Bestatin, a stimulator of polysome assembly in T cell lymphoma (L5178y). 1981; 30: 3375-77.

60. Orning L, Krivi G, Bild G et al. Inhibition of leukotriene A_4 hydrolase/aminopeptidase by captopril. J Biol Chem 1991; 266: 16507-11.

61. Minami M, Ohishi N, Mutoh H et al. Leukotriene A_4 hydrolase is a zinc-containing aminopeptidase. Biochem Biophys Res Commun 1990; 173:620-6.

62. Izumi T, Minami M, Ohishi N et al. Site-directed mutagenesis of leukotriene A4 hydrolase and aminopeptidase activities. J Lipid Mediators 1993; 6:53-8.

63. Simmons WH, Orawski AT. Membrane-bound aminopeptidase P from bovine lung. Its purification, properties, and degradation of bradykinin. J Biol Chem 1992; 267:4897-903.

64. Orawski AT, Simmons WH. Purification and properties of membrane-bound aminopeptidase P from rat lung. Biochemistry 1995; 34:11227-36.

65. Yoneda J, Saiki I, Fujii H et al. Inhibition of tumor invasion and extracellular matrix degradation by ubenimex (bestatin). Clin Exp Metastasis 1992; 10:49-59.

66. Fujii H, Nakajima M, Saiki I et al. Human melanoma invasion and metastasis enhancement by high expression of aminopeptidase N/CD13. Clin Exp Metastasis 1995; 13:337-44.

67. Saiki I, Fujii H, Yoneda J et al. Role of aminopeptidase N (CD13) in tumor-cell invasion and extracellular matrix degradation. Int J Cancer 1993; 54:137-43.

68. Delmas B, Gelfi J, Kut E et al. Determinants essential for the transmissible gastroenteritis virus-receptor interaction reside within a domain of aminopeptidase-N that is distinct from the enzymatic site. J Virol 1994; 68:5216-24.

69. Taylor WL, Dixon JE. Characterization of a pyroglutamate aminopeptidase from rat serum that degrades thyrotropin-releasing hormone. J Biol Chem 1978; 253:6934-40.

70. Nyberg F, Thornwall M, Hetta J. Aminopeptidase in human CSF which degrades delta-sleep inducing peptide (DSIP). Biochem Biophys Res Commun 1990; 167:1256-62.

71. Kunz J, Krause D, Kremer M et al. The 140 kDa protein of blood-brain barrier-associated pericytes is identical to aminopeptidase N. J Neurochem 1994; 62:2375-86.

72. Chauvel EN, Coric P, Llorens-Cortes C et al. Investigation of the

active site of aminopeptidase A using a series of new thiol-containing inhibitors. J Medicinal Chem 1994; 37:1339-46.

73. McCulloch R, Burke ME, Sherratt DJ. Peptidase activity of *Eschericia coli* aminopeptidase A is not required for its role in Xer site-specific recombination. Mol Microbiol 1994; 12:241-51.

74. Yeager CL, Ashmun RA, Williams RK et al. Human aminopeptidase N is a receptor for human coronavirus 229E. Nature 1992; 357:420-2.

75. Delmas B, Gelfi JL, Haridon R et al. Aminopeptidase N is a major receptor for the enteropathogenic coronavirus TGEV. Nature 1992; 357:417-20.

76. Shang F, Taylor A. Oxidative stress and recovery from oxidative stress are associated with altered ubiquitin conjugating and proteolytic activities in bovine lens epithelial cells. Biochem J 1995; 307:297-303.

77. Hough R, Pratt G, Rechsteiner M. Purification of two high molecular weight proteases from rabbit reticulocyte lysate. J Biol Chem 1987; 262:8303-13.

78. Huang LL, Shang F, Taylor A. Degradation of differentially oxidized α-crystallins in bovine lens epithelial cells. Exp Eye Res 1995; 61:45-54.

79. Taylor A. Oxidative stress and antioxidant function in relation to risk for cataract. In: Sies H, ed. Antioxidants in Disease Mechanisms and Therapeutic Strategies (A Volume of Advances in Pharmacology Series). San Diego: Academic Press, in press.

80. Johnson ES, Ma PCM, Ota IM et al. A proteolytic pathway that recognizes ubiquitin as a degradation signal. J Biol Chem 1995; 270:17442-56.

81. Ganesh S, Klingel S, Kahle H et al. Flow cytometric determination of aminopeptidase activities in viable cells using fluorogenic rhodamine 110 substrates. Cytometry 1995; 20:334-40.

82. Godin D, Marceau F, Beaule C et al. Aminopeptidase modulation of the pharmacological responses to synthetic thrombin receptor agonists. Eur J Pharmacol 1994; 253:225-30.

83. Kim J, Kim K. The use of FAB mass spectrometry and pyroglutamate aminopeptidase digestion for the structure determination of pyroglutamate in modified peptides. Biochem Mol Biol Int 1995; 35:803-811.

84. Klebert S, Kratzin HD, Zimmermann B. Primary structure of the murine monoclonal IgG2a antibody mAb735 against alpha (2-8) polysialic acid. 2. Amino acid sequence of the heavy (H-) chain Fd' region. Biol Chem Hoppe-Seyler 1993; 374:993-1000.

85. Yoshpe-Besancon I, Gripon JC, Ribadeau-Dumas B. Xaa-Pro-dipeptidyl-aminopeptidase from *Lactococcus lactis* catalyses kinetically controlled synthesis of peptide bonds involving proline. Biotech Appl Biochem 1994; 20:131-40.

86. Yoshpe-Besancon I, Auriol D, Monsan P et al. Reversible enzymic protection of the alpha-amino group of amino acid derivatives using an aminopeptidase A. Biotechnol Applied Biochem 1993; 18(pt 1):93-102.

87. Offner GD, Gong D, Afdhal NH. Identification of a 130-kilodalton human biliary concanavalin A binding protein as aminopeptidase N. Gastroenterol 1994; 106:755-62.

88. Eisenhauer DA, Berger JJ, Peltier CZ et al. Protease activities in cultured beef lens epithelial cells peak and then decline upon progressive passage. Exp Eye Res 1988; 46:579-90.

89. Harris CA, Hunte-McDonough B, Krauss MR et al. Induction of leucine aminopeptidase by interferon gamma identification by protein microsequencing after purification by preparative two-dimensional gel electrophoresis. J Biol Chem 1992; 267:6865-9.

90. Shapiro LH, Ashmun RA, Roberts WM et al. Separate promoters control transcription of the human aminopeptidase N gene in myeloid and intestinal epithelial cells. J Biol Chem 1991; 266: 11999-12007.

91. Keller SR, Scott HM, Mastick CC et al. Cloning and characterization of a novel insulin-regulated membrane aminopeptidase from Glut4 vesicles. J Biol Chem 1995; 270:23612-8.

92. Knight PJ, Knowles BH, Ellar DJ. Molecular cloning of an insect aminopeptidase N that serves as a receptor for *Bacillus thuringiensis* CryIA(c) toxin. J Biol Chem 1995; 270:17765-70.

93. Conlin CA, Hakensson K, Liljas et al. Cloning and nucleotide sequence of the cyclic AMP receptor protein-regulated *Salmonella typhimurium pepE gene* and crystallization of its product, an a-aspartyl dipeptidase. J Bacteriol 1994; 176:166-72.

94. Ludewig M, Fricke B, Aurich H. Leucine aminopeptidase in intracytoplasmic membranes of *Acinetobacter calcoaceticus*. J Basic Microbiol 1987; 27:557-63.

95. Aurich H, Jahreis G, Fricke B et al. Characterization of a bacterial aminopeptidase bound to intracytoplasmic membranes. Biol Chem Hoppe-Seyler 1986; 367:175.

96. Carter TH, Miller CG. Aspartate-specific peptidase in *Salmonella typhimuriam:* mutants deficient in peptidase E. J Bacteriol 1984; 159:453-9.

97. Mayo B, Kok J, Venema K et al. Molecular cloning and sequence analysis of the X-prolyl dipeptidyl aminopeptidase gene from *Lactococcus lactis* subsp. *cremoris*. Appl Environ Microbiol 1991; 57:38-44.

98. Yan T-R, Ho S-C, Hou C-L. Catalytic properties of X-prolyl dipeptidyl aminopeptidase from *Lactococcus lactis* subsp. *cremoris* nTR. Biosci Biotech Biochem 1992; 56:704-7.

99. Vesanto E, Varmanen P, Steele JL et al. Characterization and expression of the *Lactobacillus helveticus* pepC gene encoding a gen-

eral aminopeptidase. Eur J Biochem 1994; 224:991-7.

100. Avora G, Lee BH. Purification and characterization of an aminopeptidase from *Lactobacillus casei* subsp. *rhamnosus* 593. Biotechnol Appl Biochem 1994; 19:179-92.

101. Aphale JS, Strohl WR. Purification and properties of an extracellular aminopeptidase from *Streptomyces lividans* 1326. J Gen Microbiol 1993; 139:417-24.

102. Kiefer-Partsch B, Bockelmann W, Geis A. Purification of an X-prolyl dipeptidyl aminopeptidase from the cell wall proteolytic system of *Lactococcus lactis* subs. *cremoris*. Appl Microbiol Biotechnol 1989; 31:75-8.

103. Bordallo J, Cueva R, Suarez-Rendueles P. Transcriptional regulation of the yeast vacuolar aminopeptidase yscI encoding gene (APE1) by carbon sources (published erratum) FEBS Lett 1995; 369:353.

104. Frey J, Rohm KH. Subcellular localization and levels of aminopeptidases and dipeptidases. Biochim Biophys Acta 1978; 527:31-41.

105. Jones EW, Zubenko GS, Parker RR. *Pep4* gene function is required for expression of several vacuolar hydrolases in *Saccharomyces cerevisiae*. Genetics 1982; 102:665-77.

106. Roncari G, Zuber H. Thermophilic aminopeptidase: API from *Bacillus stearothermophilus*. Meth Enzymol 1970; 19:544-52.

107. Moser P, Roncari G, Zuber H. Thermophilic aminopeptidases from *Bacillus stearothermophilus*. Int J Protein Res 1970; 2:191-207.

108. Roncari G, Stoll E, Zuber H. Thermophilic aminopeptidase I; Meth Enzymol 1976; 45 B:522-30.

109. Miller CG, Miller JL, Bagga DA. Cloning and nucleotide sequence of the anaerobically regulated *pepT* gene of *Salmonella typhimurium*. J Bacteriol 1991; 173:3554-8.

110 Lazduski C, Busuttil J, Lazdunski A. Purification and properties of a periplasmic aminoendopeptidase from *Eschericia coli*. Eur J Biochem 1975; 60:363-9.

111. Gharbi S, Belaich A, Murgier M et al. Multiple controls exerted on *in vivo* expression of the *pepN* gene in *Escherichia coli*. Studies with *pepN-lacZ* operon and protein fusion strains. J Bacteriol 1985; 163:1191-5.

112. Luzdunski A, Pellessier C, Lazdunski C. Regulation of *Escherichia coli* K10 aminoendopeptidase synthesis. Eur J Biochem 1975; 60:357-62.

113. Foglino M, Lazdunski A. Deletion analysis of the promoter region of the *Escherichia coli pepN* gene, a gene subject *in vivo* to multiple global controls. Mol Gen Genet 1987; 210:523-7.

114. Strauch KL, Lenk JB, Gamble BL. Oxygen regulation in *Salmonella typhimurium*. J Bacteriol 1985; 161:673-80.

115. Murgier M, Pellissier C, Lazdunski A. Existence, localization and regulation of the biosynthesis of aminoendopeptidase in gram-negative bacteria. Eur J Biochem 1976; 65:517-20.

116. Couton JM, Sarath G, Wagner FW. Purification and characterization of a soybean cotyledon aminopeptidase. Plant Sci 1991; 75:9-17.
117. Brown SB, Krause D, Ellem KA. Low fluences of ultraviolet irradiation stimulate HeLa cell surface aminopeptidase and candidate "TGF alpha ase" activity. J Cell Biochem 1993; 51:102-15.
118. Ben Meir D, Blumberg S. In: Biologicals from Recombinant Microorganisms and Animal Cells Production and Recovery. VCH Publishers, 1991.

CHAPTER 2

STRUCTURE AND FUNCTION OF BOVINE LENS AMINOPEPTIDASE AND COMPARISON WITH HOMOLOGOUS AMINOPEPTIDASES

Allen Taylor, David Sanford and Thomas Nowell

INTRODUCTION

Lens leucine aminopeptidase (LAP[†]) is the aminopeptidase for which structural, kinetic, and mechanistic information is most developed. Crystallographic, electron micrographic, NMR, and photoaffinity labeling and modeling studies indicate that lens LAP protomers are bilobal, and that bestatin and substrates are bound in an active site, which is found in the larger lobe of each protomer. Zn^{2+} is involved in substrate liganding and presumably in catalysis of hydrolysis. There is no evidence of an acyl-enzyme intermediate

---

[†] *Abbreviations in this chapter are as follows: AP, aminopeptidase; b, bovine; h, hog; l, lens; k, kidney; LAP, leucine aminopeptidase; LeuP, leucinephosphonic acid; MPD, methyl pentanediol; Me^{2+}, generic divalent metal ion; kDa, kilodalton*

Aminopeptidases, edited by Allen Taylor. © 1996 R.G. Landes Company.

in hydrolysis. Amino acid sequences determined directly or deduced from cDNAs indicate homologies between lens aminopeptidase and APs in organisms as diverse as *E. coli*, particularly in catalytically important residues, or in residues involved in metal ion binding. For additional details regarding structures and function of some other APs, readers are referred to subsequent chapters in this book.

COMPOSITION AND STRUCTURE

Like several of the other APs, bovine lens (bl) LAP is synthesized as a larger precursor of 514 amino acids[1] (Table 2.1). This is reduced to an oligomer containing 487 amino acid residues and 2 (or possibly 3) Zn^{2+}[1,2] (see section on metal ions). Although crystals of LAP were obtained over two decades ago, limitations in technology thwarted their use until the last decade. Initial information regarding the composition and structure of the lens LAP molecule and the organization of the blLAP hexamer were obtained by biochemistry,[3-6] and electron microscopy of single molecules and crystal thin sections[7,8] (Fig. 2.1a). The shape of the blLAP protomer is bilobal (Fig. 2.1b-e); the distribution of protein being 2/3, 1/3 between the larger and smaller lobes, respectively.[7,9] Hexamers appear as two concentric triangles, the smaller being offset from the larger "less dense" triangle by 60° (Fig. 2.1f). The shape of the protomer and the arrangement of the identical protomers within the hexamer (a dimer of trimers) were corroborated by crystallographic analysis.[8,10] Aspartate carbonyl tansferase has identical appearance in the electron microscope. However, despite the identical appearance of protomers and assemblies of blLAP, we found no AP activity in purified aspartate transcarbamylase and vice versa.[7,8] Centers of each of the two blLAP hexamers contained in the lens LAP unit cell were at (1/3, 2/3, 1/4) and (2/3, 1/3, 3/4) along a, b, and c axes, respectively, of the unit cell.[7] The space group for lens LAP is $P6_322$ for crystals grown in ammonium sulfate, MPD, or lithium sulfate[7,8,10] (Taylor, unpublished).

The maximum length of the blLAP protomer along the molecular 3-fold axis of symmetry is 90 Å. The edge of the triangle is ≈113 Å[7,10] (Fig. 2.1a). The hexagonal needle crystals have unit cell parameters a = b = 132 Å, c = 122 Å.[8] The same parameters describe bovine pancreas LAP.[11]

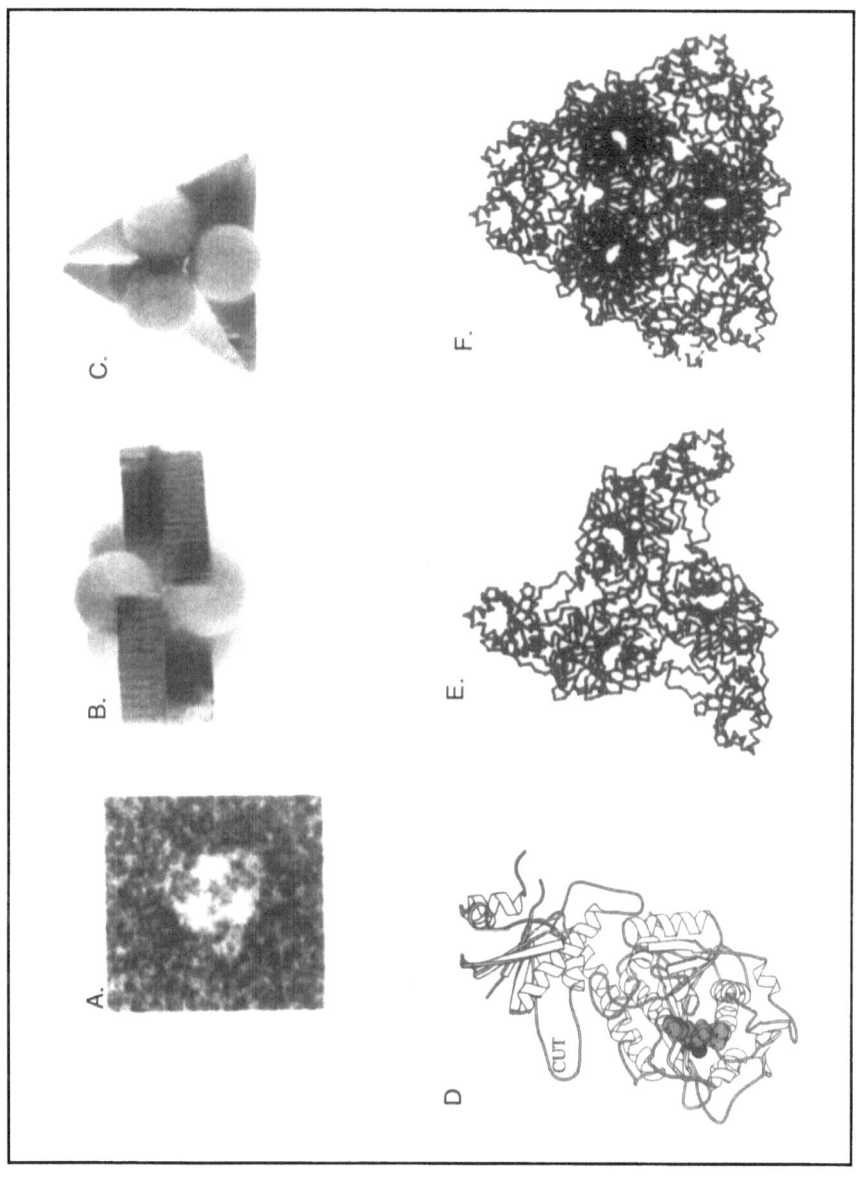

Fig. 2.1. Structure of bovine lens LAP. (a) Enhanced electron micrograph of a single blLAP molecule embedded in sodium phosphotungstate. (b and c) Photographs of LAP hexamer model along the 2-fold (b) and 3-fold (c) axis of symmetry. The spherical portion of the subunit contains about 2/3 of the protein and the smaller appendage (shown here as a wedge) contains about 1/3 of the protein. (d) Schematic ribbon drawing of the monomer based on crystallographic data.[10] Trypsinization of the enzyme results in cleavage at a unique site, indicated on figure as CUT (between residues 137 and 138). Light spheres show the bestatin binding site. The zinc ions are behind the bestatin and are shown as dark spheres. (e and f) Tracing of the carbon backbone of the blLAP trimer (e) and hexamer (f) as viewed down the threefold axis.

Table 2.1. Aminopeptidases for which structural and compositional data are available

Name[a]	Source	Subunit mass (kDa) [# of residues in mature form of protomer]	# of subunits	Total mass (kDa)	Metal ion content	Residues used to ligand metal ion or active site residues	Unique gene?	Preprotein size (kDa) (Size of amino-terminal extension kDa)	Homology with[e]	Refs.
Zinc aminopeptidase	Rat kidney membrane	140	1	140	Zn²⁺		Yes		77% to AP-M; 31% to leukotriene A₄ hydrase; 18% to E. Coli AP-N	15
Aminopeptidase M	Rat kidney membrane	110 [966]			Zn²⁺	His, His, Glu				74
Arginine aminopeptidase	Liver	90	1	90						45
Leucine aminopeptidase with	Bovine lens and 1 kidney; hog lens and kidney[c]	53 [487 res]	6	318	2 Zn²⁺ per subunit <3Å apart. #1 (488) readily exchangeable. Mg²⁺ and Mn²⁺ can bind in site 1, Co²⁺ can bind in both sites. #2 (489) tight binding	Asp,Asp,Glu; Asp, Asp,Glu, Lys	Yes	55.6 (bl,bk[d]) 514 res preprotein (26 residue extension)	arabidopsis LAP; prolyl AP; 31/52% to AP-A; xerB gene product; Pep A; aeromonas LAP	42% 3 9 10 14 20 22 31 39 40 49 54 82 102

Continued...

Enzyme	Source	Mr (kDa) [residues]	Subunits	Mr (holoenzyme)	Metal	Metal ligands	Homology to bILAP	Notes	References
Aminopeptidase A *xerB* gene product	*E. coli*	55.3 [503]					see bILAP	bILAP; *S. Typhimurium* pepA; 49% identity with arabidopsis AP	50, 70
Aminopeptidase A (*pep4* gene product)	*S. cerevisiae*	44			2 Zn²⁺		Yes		84
ysc1, APEI/LAP4	*Saccharomyces cerevisiae*	44.8 [469]; 51.7	12	640			Yes	57.0 (514 residues); 57.2 (514 residues)	16, 85
Aminopeptidase	*Aeromonas*	30, 32	1		2 Zn²⁺ 3.5A apart; 1 Zn²⁺ required, can bind another Zn²⁺; cocatalytic role for both Zn²⁺	Asp, His Glu		43 (≈13 kDA); structural homology but little primary sequence homology with bILAP	17, 51, 52, 86
Dipeptidase	Mouse ascites				1 Zn²⁺				29
Methionine aminopeptidase	Human liver	[478]			2 Co²⁺			p67 and *E. coli* methionine aminopeptidase	87
Methionine aminopeptidase	*Salmonella*	34			no metal				88
Methionine aminopeptidase	*E. coli*	29.3 [264]	1	29.3	2 Co²⁺	Asp, Asp, Glu, His		19% with N-terminus of pseudomonas creatinase; aminopeptidase P prolidase	53, 76, 89

Continued...

Table 2.1. Aminopeptidases for which structural and compositional data are available (continued)

Name[a]	Source	Subunit mass (kDa) [# of residues in mature form of protomer]	# of subunits	Total mass (kDa)	Metal ion content	Residues used to ligand metal ion or active site residues	Unique gene?	Preprotein size (kDa) (Size of amino-terminal extension kDa)	Homology with[c]	Refs.
Methionine aminopeptidase	S. Cerevisiae	[387, 377 mature]			Co^{2+}		Yes		40% Met AP from E. Coli, S. Typhimurium, B. Subtilis	18 90
				≈42 ≈34	Co^{2+}					
Methionine aminopeptidase	Hog liver	≈70	1	≈70	Co^{2+}			≈70		91
D-amino acid aminopeptidase	Ochrobactrum anthropi SCRC C1-38	59	2		no metal					92
Peptidase C	Streptococcus thermophilus CNRZ302	445 50.4	n=6			Cys			70% to PepC from lactococcus luctis 38% to bleomycin hydrolase	77 93
Peptidase C[b]	Lactobacillus helviticus	51.4				Cys			48% to Pep C from lactococcus lacti 98% to lactococcus helviticus CNRZ32	78

Continued...

Enzyme	Source	Subunit MW	Native MW	Metal/subunits	Active site residues	Homology	Ref
Bleomycin hydrolase	*Saccharomyces cerevisiae* Nonvacuolar	55.4	220	n=4		100% w. LAP3 see peptidase C	93
Aminopeptidase Nb	Muscle	51	390				94
Aminopeptidase N	Cerebral pericytes	140				100% to 140 blood-brain barrier protein	95
Aminopeptidase N					Val,X,X,His, Glu,X,X,His		96
Aminopeptidase P (aminoacyl-prolylpeptide hydrolase E.C. 3.4.11.9)	Rat lung membrane	90	220-340 glycoprotein				97
	Bovine lung	95	360 glycosyl-phosphatidyl-inositol anchor				98
Prolidase	Hog kidney	55	110	Co^{2+} 2	Asp,Asp,His, Glu, Glu, based on Met AP sequence	Aminopeptidase P, MetAP from *E. coli*	99

[a] In some cases, similar enzyme names are used, but it is not clear that the enzymes are related. Thus they are listed separately. Where clear associations have been noted, the enzymes are grouped together.

[b] Not inhibited by bestatin.

[c] Bovine lens and hk may differ in that activity is detected in hkLAP in which only 1 equivalent of Zn^{2+} is bound per subunit.

[d] bk=bovine kidney; bl=bovine lens

[e] This is only a partial list. See text for further details regarding the homologies indicated.

When NaCl was added to the MPD, crystals with the space group P321 and unit cell dimensions a = 130.4 Å, c = 125.4 Å were obtained[2] for blLAP cocrystallized with LeuP (leucinephosphonic acid). Although approximately half of the interhexamer interactions are different between crystal lattices in the P321 and $P6_322$, when only one subunit is superimposed there is a calculated rms deviation of only 0.36 Å between the structures for the native and blLAP-LeuP complex. Thus, there is no significant structural change induced to the enzyme upon binding of LeuP. The same pertains to binding of other inhibitors including bestatin, [(2S,3R)-3-amino-2-hydroxy-4-phenylbutanoyl]-L-leucine, and amastatin, [(2S,3R)-3-amino-2-hydroxy-5-methylhexanoyl]-L-valyl-L-valyl-aspartic acid,[12] as well (see the next section).

The shape of blLAP protomers and their arrangement within the highly homologous porcine kidney (hk[13]) LAP hexamer is indistinguishable from that described for the bovine enzyme, although hkLAP crystallized in the space group $P2_12_12_1$.[14] Unit cell parameters for hkLAP are: a = 186 Å, b = 223 Å, and c = 80 Å.

The availability of a full-length clone and a deduced amino acid sequence for a purportedly identical LAP from bovine kidney (bk) LAP[1] (see Homologies below) indicated that the prior lens LAP sequence obtained by protein sequencing techniques could be improved.[6,13] A major difference is that LAP is apparently synthesized with a 26 amino acid extension in kidney and presumably in other tissues as well.[1] Other aminopeptidases also show prosequences which include amino-terminal extensions[15-18] (Table 2.1). An octapeptide (473-481) that was probably overlooked because it is a trypsin fragment is also indicated. Three internal differences were also found. The kidney sequence has a Ser at position 45 instead of Pro as indicated in the lens sequence, a Leu inserted after Trp at position 382, and a Met replacing the Trp at position 383. All these assignments were required to fit the crystallographic data for crystals of blLAP. This information also provides further confirmation of immunological data which indicate that LAP from different organs within a species are identical.[13] The mass of LAP is 52,989Da, whereas the mass of the proprotein would be 55,562Da.[1]

Trypsinization of the enzyme resulted in formation of only two polypeptides, one with a mass of 37 kDa (residues 138-487) and the other with a mass of 18 kDa (residues 1-137) (Fig. 2.1d). A unique cleavage between residues 322 and 323 by hydroxylamine results in similarly sized but different polypeptides (1-322 and 323-487) from those obtained with trypsin.[6]

The NH$_2$-terminal 150 residues fold to give a B-sheet sandwiched between four alpha-helices. The N-terminus itself occurs as the middle strand of the sheet.[10] It is this portion of the molecule which comprises the smaller lobe of each protomer. There is a long loop connecting an alpha-helix and the fifth strand of the beta sheet. This long loop contains the peptide bond between residues Arg-137 and Lys-138, where trypsin cleaves each protomer within the hexameric enzyme but leaves the hexamer intact, fully active,[19] and crystallizable (Carpenter, unpublished). Monomers of blLAP have not been isolated. The effective stabilization of hexamers in the native and, presumably, in the trypsin-treated enzyme is achieved via an extensive network of hydrogen bonds and van der Waals' interactions.[2,10,20]

Assembly of blLAP protomers results in a solvent cavity near the center of hexamers. The cavity is ≈15 Å in radius and 10 Å in height along the 3-fold axis.[10] Solvent channels which run along the molecular 2-fold axes give access to the central solvent cavity. The dimensions of the solvent cavity would appear to be consistent with the preference of AP for di- and tripeptides but are inconsistent with the ability of blLAP to hydrolyze lens α-crystallin, a 20 kDa protein.[21] Despite the multimeric structure of many aminopeptidases, kinetic advantages, i.e., allostery, cooperativity, etc. of a polymeric structure have rarely been described.[21a]

INHIBITOR BINDING SITE

Before proposing a mechanism of inhibitor binding or a mechanism of peptide hydrolysis, it was necessary to determine the number of active sites per hexamer, locate the inhibitor and/or substrate binding sites, and reconcile previously observed metal-ion/substrate interactions. The discovery of a relatively tight-binding transition state inhibitor of LAP, bestatin, provided new opportunities to characterize the active site[12] (for review, see refs. 9, 22).

BESTATIN IS A SLOW, RELATIVELY TIGHT-BINDING, COMPETITIVE INHIBITOR OF BLLAP

In order to begin to use the available structural data to propose a mechanism of action, it was necessary to determine the type of inhibition of blLAP by bestatin. Three approaches were used to do this: (1) enzyme was added to a solution containing

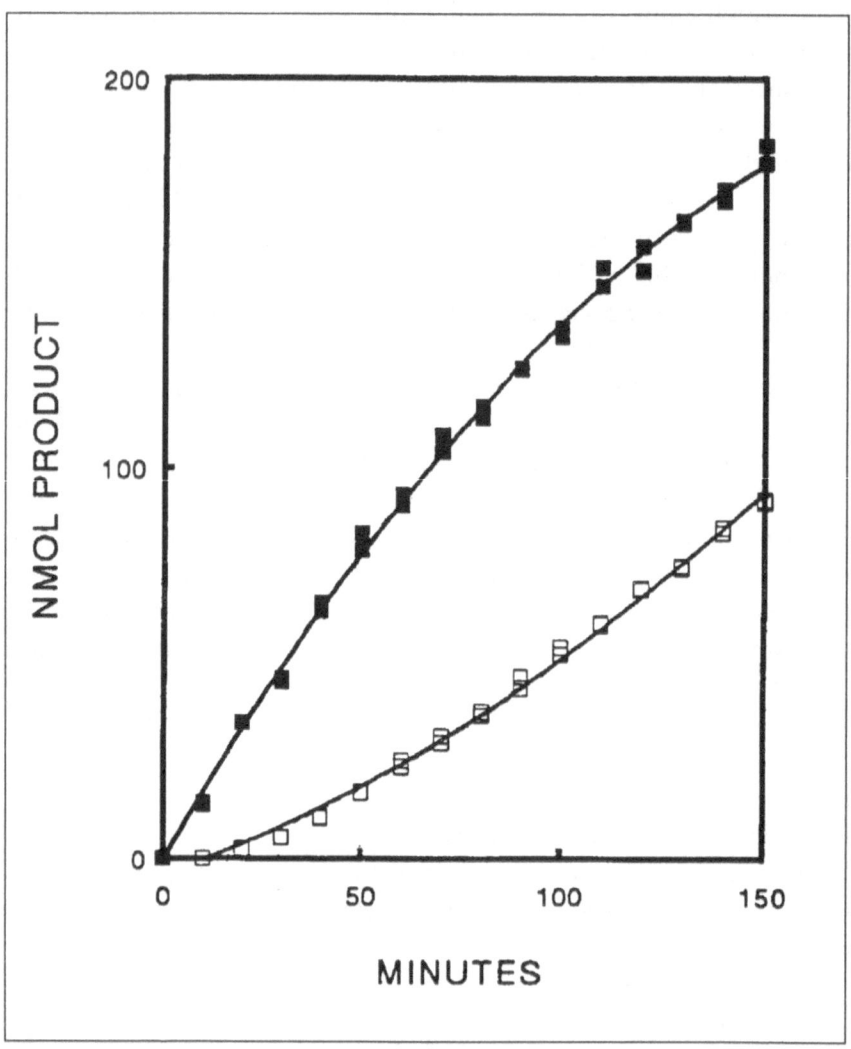

Fig. 2.2. Association and dissociation of bestatin with blLAP.[9] Upper curve: blLAP 1.7x10⁻⁹ M subunits (final concentration) was added to a solution of 9.5 mM LeuGlyGly and 10⁻⁸ M bestatin (final concentrations), and samples tested for activity at the times indicated. Lower curve: blLAP was incubated with bestatin for 20 min. This solution was then added to a solution of substrate. Activity determinations and the final concentration of bestatin were as indicated for upper curve.

LeuGlyGly (a physiologically relevant substrate[23-25]) and bestatin (Fig. 2.2, upper curve);[9,22] (2) a solution containing enzyme and inhibitor was added to a solution containing substrate and inhibitor such that there was no change in the concentration of inhibitor (data not shown); and (3) a solution containing enzyme and inhibitor was added to substrate (Fig. 2.2, lower line). In all cases the final concentration of inhibitor was the same.

When blLAP was added to solutions of LeuGlyGly containing bestatin, the initial rapid rate of evolution of product decreased by 71% after approximately 110 min (Fig. 2.2, upper curve). The $t_{1/2}$ for establishing steady-state levels of the EI* complex was approximately 30 min under these conditions. As expected, when the bestatin/blLAP ratio increased, $t_{1/2}$ was reduced. The reduced activity was not due to substrate depletion. The same final velocity is observed in assays in which LAP was preincubated with an excess of bestatin and then added to the substrate (Fig. 2.2, lower curve). In the latter instance the reaction progress curve turns upward as the high inhibitor concentration initially present in the preincubation mixture is diminished. The regain in activity shown by the lower curve indicates that the inhibition of blLAP by bestatin is reversible. The reaction can be described by the following scheme:

$$
\begin{array}{ccccc}
E & + & S & \rightleftarrows & ES \rightarrow E + P \\
+ & & & & k_{on} \qquad\qquad \text{Scheme 1}\\
 & & k_3 & & k_5 \\
I & & \rightleftarrows & EI & \rightleftarrows \quad EI^* \\
 & & k_4 & & k_6 \\
 & & & k_{off}
\end{array}
$$

The $t_{1/2}$ for the deformation of the tightly bound complex, EI*, is approximately 22 min. Comparable data were obtained with azidobestatin.[22] The similarity of the rate constants obtained when [I]/[E] ratios were 10 or 100 provides justification for use of kinetic analysis for a slow binding inhibitor.[26]

The dissociation constant for the initial (EI) collision complex is $K_i = k_3/k_4$, and the dissociation constant, K_i^*, for the final complex, EI*, is $k_{off}/k_{on} = k_4k_6/k_3k_5$.[26-28] k_{on} and k_{off} are the rate constants for formation and deformation of EI*. K_i^* can be calculated either from the ratio of k_{off}/k_{on} as obtained from presteady-state experiments or determined directly using data from both phases of the reaction. The apparent first order rate constant, λ, and the apparent second order rate constant for formation of the enzyme-inhibitor complex, β, were determined according to Cha[26] and replotted in a reciprocal form, $1/\beta = 1/k_{on} + S/k_{on}K_m$. The linearity of the plot indicates that bestatin is a competitive inhibitor.

Knowledge that bestatin is a slow-binding inhibitor of blLAP and that bestatin competes with substrate for the same binding site on blLAP, allowed use of presteady and steady-state data to further describe the binding process.

For competitive inhibitors it is also possible to derive rate and binding constants for the major steps of the reaction pathway shown in Scheme 1 using presteady-state and steady-state portions of the progress curves (Equations 1 and 2).[27]

$$v_s = V_{max}S/K_m(1+I/K_i^*)+S \qquad (1)$$
$$v_o = V_{max}S/K_m(1+I/K_i)+S \qquad (2)$$

K_i and K_i^*, respectively, are 1.1×10^{-7} M and 1.3×10^{-9} M. These data indicate that bestatin and LAP are bound approximately 84-fold more tightly in the final EI* complex than in the initial collision complex. The apparent rate constant is $k = 3.4 \times 10^{-4}$ sec^{-1}. The rate constant for the formation of the final complex is obtained from $k_5 = k_6(K_i/K_i^*-1) = 1.5 \times 10^{-2}$ sec^{-1}. This is in reasonable agreement with the value of k_5 calculated from k_{on} [2.38 ($\pm$0.49) $\times 10^3$ sec^{-1} M^{-1}] determined from presteady-state data and the diffusion controlled rate (10^7 M^{-1}sec^{-1}). The rate constant for deformation of the final complex from the initial collision complex is obtained from the expression $k_6 = kv_s/v_0 = 2 \times 10^{-4}$ sec^{-1}. Finally, the rate constant for the dissociation of the collision complex, k_4 (1 sec^{-1}), is obtained from $K_i = k_4/10^7$ M^{-1}s^{-1}. This is corroborated by a value for k_{off} of 8.29 ($\pm$1.4) $\times 10^{-5}$ sec^{-1}. These data indicate that slow achievement of a steady-state involves slow deformation of the EI*. Thus, the slow binding of bestatin involves rapid formation of the initial collision complex (EI), slow transformation of the EI to the tight complex EI*, and even slower deformation of that complex.[9]

Patterson[29] suggested that slow binding may be due to slow removal or displacement of an OH$^-$ from a "deep pocket" in an

Table 2.2. Inhibition constants for bestatin and aminopeptidases

Enzyme	Type of inhibitor	K_i^* (M) [substrate][a]	K_i (M)	Me^{2+}	Refs.
Bovine lens leucine amino peptidase	compet, slow	1.3×10^{-9} [LeuGlyGly]	1.1×10^{-7}	Zn^{2+}, Zn^{2+}	21-23
Hog kidney cytosolic leucine aminopeptidase	compet, slow	2×10^{-8} [LpNA]		$Zn^{2+}Mg^{2+}$	32
	compet, slow	2×10^{-8} [LpNA]		$Zn^{2+}Mg^{2+}$	61
	compet, slow	5.8×10^{-10} [LpNA]		$Zn^{2+}Mn^{2+}$	33
	compet, slow	2×10^{-8} [LeuGlyGly]	$\approx 10^{-8}$	$Zn^{2+}Zn^{2+}$	
		2×10^{-8} [LβNA]		$Zn^{2+}Zn^{2+}$	12
(Thiobestatin) (bestatin thiamide)		5.5×10^{-7} 3.3×10^{-7}			61 61
Bovine lens leucine aminopeptidase (azidobestatin)[b]	compet, slow	4×10^{-9} [LeuGlyGly]	10^{-8}	Zn^{2+}, Zn^{2+}	22
Aeromonas	compet, slow	1.8×10^{-8} [LpNA]			33
Aminopeptidase M	compet, slow	4.1×10^{-6} [LpNA]	7×10^{-6}	$Zn^{2+}Mg^{2+}$	32 61
	not slow	1.4×10^{-6} [LpNA]			33
Aminopeptidase B	compet, slow	6×10^{-8} [LβNA]			100
Mouse ascites tumor dipeptidase	compet, slow	2.7×10^{-9} [dipeptides]		Mg^{2+}	29
Aminopeptidase A (amastatin)		2.5×10^{-7}		Zn^{2+}	68

[a] The K_m for all the substrates is approximately mM.
[b] Note the inhibitor used was azidobestatin.

aminodipeptidase. It is possible that the C2-OH in bestatin (which is involved in the tighter- and slow-binding) displaces the putative OH⁻ in a two step process involving the initial and final complexes[9] (Table 2.2). Kim et al used this and metal binding data to propose that slow binding involves initial complex formation between the P_1 hydroxyl and Zn^{2+}488.[30] Subsequent dissociation of this complex yields a complex in which the hydroxyl and the αNH_2 of bestatin is bound to Zn^{2+}489.[10,30,31]

COMPARISON OF BINDING CONSTANTS FOR BESTATIN AND LAP WITH CONSTANTS FOR BINDING OF BESTATIN TO OTHER AMINOPEPTIDASES[9,22]

The K_i^* value for binding of bestatin to blLAP falls within the range of K_i^* of 5.8 x 10⁻¹⁰ and 4.1 x 10⁻⁶ for bestatin and all other bestatin-inhibited APs studied to date (Table 2.2). Further comparison of binding of bestatin to APs was initially made difficult since the K_i^* noted by Rich for hkLAP is approximately 34-fold higher than that noted by Wilkes and Prescott for the same enzyme.[32,33] The reasons for these discrepancies become clearer when the equations which are used to calculate K_i^* are recast (equations 1b and 2b) to emphasize relationships between K_i and V/v_o, or K_i^* and V/v_s,

$$K_i^* = K_m I/[VS/v_s - K_m\text{-}S] \qquad (1b)$$
$$K_i = K_m I/[VS/v_o - K_m\text{-}S] \qquad (2b)$$

It is observed that a change of only 5-fold in the ratio of V/v results in differences in K_i or K_i^* of over 40-fold. In various laboratories differences in experimental technique (i.e., different metal ion content of the enzyme, time of assay, substrate, etc.) could result in small differences in V/v. Thus, it is plausible that APs share modes of bestatin binding which may be even more similar than suggested by the data in Table 2.2, despite their diverse apparent specificity.

STOICHIOMETRY

Many compounds were synthesized in attempts to characterize binding sites of APs (see Inhibition Binding Site). Analogs which incorporate elements of (putative) transition state intermediates between substrates and products are frequently bound more tightly than substrates to the respective enzymes. As noted in Figure 2.3, bestatin contains these putative transition state structural elements.

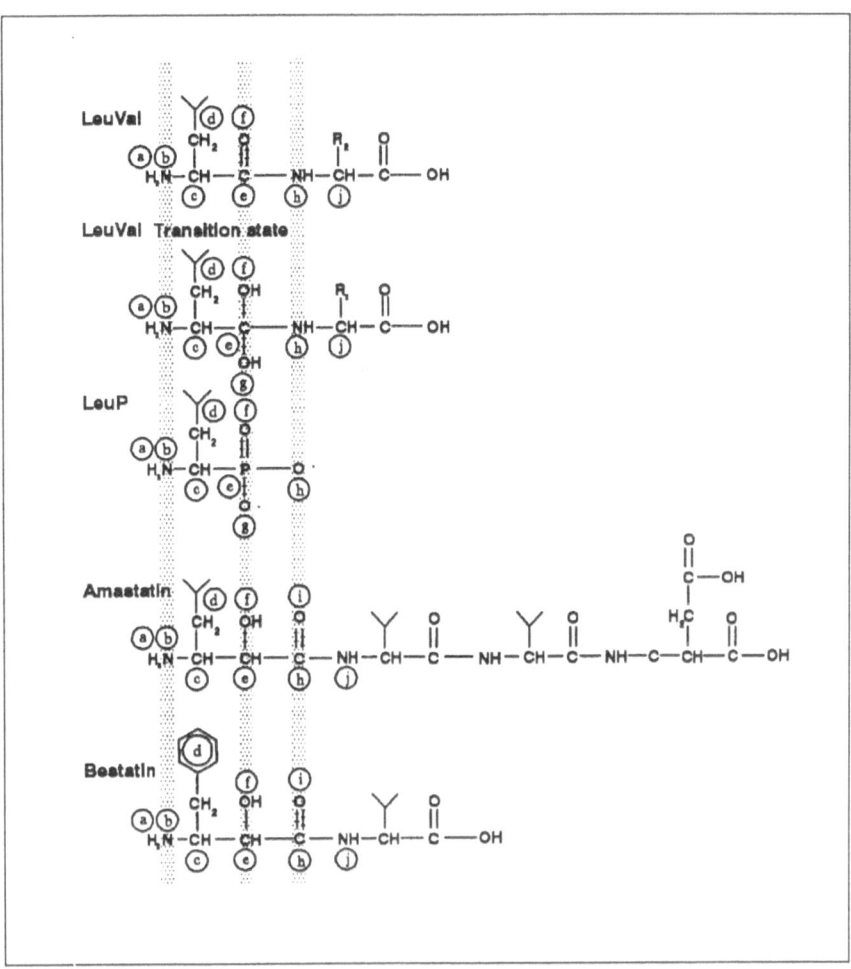

Fig. 2.3. Structures of a theoretical substrate LeuVal, the putative LeuVal transition state, leucinephosphonic acid, amastatin, and bestatin with positions of comparable functional groups indicated. The substrate PheLeu has the same functional groups as shown in bestatin other than the tetrahedral carbon labeled (e). However, the stereochemistry is reversed.

Direct binding measures indicate that six equivalents of bestatin are bound per LAP hexamer.[9] This is consistent with kinetic data which indicate that only with six equivalents of bestatin bound per hexamer is the blLAP completely inactivated.[9,34] Complete occupancy of each binding site in the enzyme-inhibitor complex has been corroborated by kinetic studies with p-azidobestatin[22] and by structural studies of blLAP complexed with bestatin,[10] amastatin[20] and leucinephosphonic acid.[2] A ratio of 1:1 bestatin/monomer was also obtained in *Aeromonas* AP[33] and yeast AP-I.[35] This information is in contrast with data which indicate that there is 80% inhibition of hkLAP when only one bestatin is bound per hexamer.[33]

IDENTIFICATION OF ACTIVE SITE AND MODE OF INHIBITOR BINDING

Compendia of kinetic data obtained with many peptides, esters, and amino acid analogs are found in prior reviews.[23,36-39] Initial attempts to label AP active sites were based on use of such peptides into which nucleophilically labile functional groups were incorporated (reviewed in refs. 22, 23, 39, 40). These included the use of chloromethyl ketones, fluoromethyl ketones, amino acid benzenesulfonyl fluorides, thiol and serine reagents, etc. The inhibition constants for these inhibitors are of similar magnitude to K_i and K_m for the substrates. The inability to label the active site with these inhibitors casts doubt on the presence of an enzyme-bound nucleophile at the active site[23,41] (see the metal ions section). These reagents did, however, yield information regarding the active site.

NMR studies of the substrate analog leucyl-o-sulfonic acid showed for the first time that substrates were within a hydration radius of Mn^{2+} in blLAP in which the "more readily exchangeable" or "activation site" Zn^{2+}488 has been replaced by Mn^{2+}.[5,10,31,41,42] Kinetic investigations had predicted comparable relationships decades earlier.[12,43,44] Perturbation of the cobalt spectrum upon binding inhibitors suggests that substrates or inhibitors are within the immediate vicinity of metal ions in *Aeromonas* AP as well.[33]

Bestatin and amastatin are relatively tight-binding inhibitors. Among the features that distinguish these "statin"-containing moieties (i.e., bestatin and amastatin) is a tetrahedral carbon atom positioned between the scissile carbonyl and the carbon which bears the $\alpha NH2$ group (Fig. 2.3). This tetrahedral carbon has a hydroxyl which mimics the presumptive tetrahedral transition state which is formed after nucleophilic addition of OH^-. The importance of this tetrahedral carbonyl is indicated by the K_i ($K_i \approx 10^{-8}$ M) of such inhibitors (Table 2.2), which is $\approx 10^{-5}$ the K_i of peptide substrates.[23] Bestatin, epibestatin, and epiamastatin, which have their C2 in the *R* configuration instead of the *S* configuration, bind to hkLAP >10^2 less tightly than do bestatin and amastatin.[32] Thus, stereochemistry at C2 is of importance in determining avidity of binding of these inhibitors. In addition, the carbonyl retains importance as indicated by an increase in Ki of 5×10^4 when the carbonyl group is changed to a methylene.[45] This is consistent with interpretation that this dramatic difference in binding is due to loss of the carbonyl interaction with Zn^{2+}. However, it should be noted that in the bestatin and amastatin cocrystal structures, the P_1 carbonyl oxygen atom is not liganded to Zn^{2+}488. The loss of binding energy has been rationalized as due to loss of the inhibitor-P_1 carbonyl-Lys-262 interaction[30] (Fig. 2.4, Table 2.3), but the long interatomic distances between the Lys-NH_2 and scissile carbonyl oxygen predicted for model substrates suggest other rationales should also be considered.

A variety of other transition state analogs have been shown to bind to LAP. Amastatin has the αNH2-terminal and a Leu side chain as part of the "statin" component, as well as an extended peptide chain (Fig. 2.3). It is bound less tightly to blLAP but more tightly to other APs.[32] L-leucinephosphonic acid mimics the putative *gem*-diolate of a putative transition state which might be found during hydrolysis of peptides, and as expected, it is bound (0.23 µM to hkLAP) more tightly than peptides but less tightly than some "statins".

The availability of a relatively tightly-bound LAP inhibitor such as bestatin and the atomic coordinates obtained from the single crystal structure determination of this transition state analog inhibitor[46] provided further details with respect to identification of the active site. Conversion of bestatin to [³H]p-azidobestatin resulted in an effective photoaffinity label for blLAP.[9,22] After equilibrating [³H]p-azidobestatin and LAP, activation of the label by exposure of the mixture to UV light resulted in p-azidobestatin-dependent inactivation of the enzyme. Digestion of the photolabeled LAP with trypsin or hydroxylamine indicated that the radioactivity was contained within the unique large tryptic fragment (residues 138-487) and the unique small hydroxylamine polypeptide (residues 323-487).[6,22] Thus, the bestatin binding site and presumably the active site are within the carboxyl third of the subunit. This corroborated prior NMR data[41] and was consistent with subsequently obtained crystallographic data.[10] Enzyme residues which are within 4 Å of the P_1 residue in the amastatin-blLAP complex are Thr-359, Asp-273, Asp-255, Gly-362, Thr-361, Leu-360, Lys-250, Met-270, Glu-334, Arg-336 and Asp-332

Table 2.3. *Interactions of selected functional groups of blLAP with other functional groups in the enzyme, inhibitors, or in modeled complexes with substrates*[a]

Functional Groups							
Enzyme	Inhibitor or Me^{2+}	Position in Fig. 2.3[b]	Bestatin Å	Amastatin[c] Å	LeuP[d] Å	LeuVal Å	LeuValTS Å
Zn488	O of P_1 OH	f	2.1	2.1			
	O of P_1 C=O	i	(3.3)[e]	(3.9)			
	O of P_1 C=O	f				2.9	
	O of *gem* diolate	f,g					2.4 (3.9)
	αNH_2 of P1	b				(3.9)	
	O_1 of Ph[f]	g			2.5		
	O_2 of Ph	f			2.1		
	Zn489		3.1 (2.9 in native)	3.3	3.4		
Zn489	αNH_2 of P1	b	2.3	2.1	2.3	2.1	2.2
	O of P1 OH	f	2.0	2.2			
	O of P1 C=O	f				3.2	
	O of *gem* diolate	f					2.7
	O_1 of Ph	g			2.2		
	O_2 of Ph	f			3.8		
	Zn488						
Asp-332 carboxylate O_2	O of P1-OH	f	(3.7)	(3.7)			
	O of P1 C=O	i	(3.2)	(3.6)			
	O_2 of Ph	f			2.9		
	Zn488		1.9 (2.1 in native blLAP)				
Asp-332 carboxylate O_1	O of P1 C=O	i	(3.3)	(3.6)	3.5		
	O_2 of Ph	f		(3.7)			
	P1' isopropyl of amastatin	d					

Residue	Partner atom						
Asp-332 carbonyl O	O of P₁-OH	f	(3.4)				
	O of P₁ C=O	f for LeuVal	(3.8)	(3.2)		3.2	
	NH of P1'	j	(3.6)	(3.2)			
	Zn488		2.2				
	O of gem diolate	f					(3.1)
	O₁ of Ph	g			3.1		
	O₂ of Ph	f			3.0		
Asp-255 carboxylate O₁	Zn488		2.5		2.1		
	Zn489			wid[g]	2.7		
	O of P1 OH	f	2.9*	wid			
	N of P1 αNH₂	b	(3.7)	2.4			
	O of P1 C=O	f for LeuVal		(3.1)	3.1	(2.6)	(2.6)
	O of gem diolate	f		(4.1)		3.5	
	O₁ of Ph	g			3.4		2.9*
	O₂ of Ph	f			3.0*		
Asp-255 carboxylate O₂	Zn488			2.9			
Glu-334 carboxylate O₁	O of P1 OH	f	(3.2)	(3.4)	3.3[h]		
	Zn489		2.1		3.1		
	O₁ of Ph	g		3.1		2.1	
Glu-334 carboxylate O₂	O of P1 OH	f	(3.1)	(3.6)	3.4		
	Zn488		2.4		2.0		
	O₁ of Ph	g			3.0		
Lys-250 εNH₂	Zn489		2.4		3.5		
	N of P1 αNH₂	b	(3.5)	(3.2)	3.5	3.6	
	O of P1 OH	f	(3.6)	(4.0)	3.3	3.7	
	O₁ of Ph	g					
Asp-273 carboxylate O₁	N of P1 αNH₂	b	(3.5)	(3.0)	(3.1)	(3.1)	
	Zn489				2.1		

Continued...

Table 2.3. Interactions of selected functional groups of bLAP with other functional groups in the enzyme, inhibitors, or in modeled complexes with substrates[a]

Enzyme	Inhibitor or Me^{2+}	Position in Fig. 2.3[b]	Bestatin Å	Amastatin[c] Å	LeuP[d] Å	LeuVal Å	LeuValTS Å
Asp-273 carboxylate O_2	"		(2.7)	(2.4)	(2.9)	(2.8)	(2.8)
Leu-360 C=O	HN of P_1'	j	3.5	3.9		3.0^h	
	O of P_1' C=O	f				3.2	(3.0)
	O of gem diolate	f			3.2		
	O_3 of P_h	h					
Lys-262 ε-NH2	O of P_1 C=O	i	3.0	2.7		(3.3)	2.9
	NH of P_1' NH	h			2.7		
	O_2 of P_h	f					
*Gly-362 backbone HN	COO of P_1'		2.9	2.8			
	C=O of P_1'						
*Gly-362 backbone CO	COO of P_3'			3.3			
Arg-336 N_1 of Guan	O in P_1 OH	f	7.0	3.8			
	N of P_1' NH	j		(3.4)			
	O of P_1 C=O	f					
	O_b of gem diolate	f				3.1	3.2
	P_1' isopropyl of amastatin	d		3.8			
Arg-336 N_2 of Guan	O of P_1 OH	f	9.0	4.0			
	O of P_1 C=O	f				3.9	
	O_a of gem diolate	f					3.6
	O_b of gem diolate	f					(4.0)

Residue	Group	Obs			
Met-454	P₁-hydrophob	d			
Ala-451	P₁-hydrophob	d	<4.0		3.7
*Gly-362	P₁-hydrophob	d	<4.0		
Thr-359	P₁-hydrophob	d	<4.0	3.6	3.7
Thr-359 C=O	αNH_2	b	(3.4)	(2.9)	
Thr-361	isobutyl of amastatin	d	<4.0	3.8	
Met-270	P₁'-hydrophob	d	<4.0	3.5	<4
Asn-330	P₁'-hydrophob				
Ala-333	P₁'-hydrophob				
Ile-421	P₁'-hydrophob				
$H_2O_{b,f}$					2.7
H_2O_c					2.7
H_2O_d					2.7

[a] Data summarized from refs. 2, 10, 30, 31, from files 1bll, 1lcp, 1lan in the protein data bank (refs 103,104) from coordinates for the blLAP complex with bestatin kindly provided by N. Strater and W.N. Lipscomb, and unpublished observations. In reference no. 101, Interatomic distances were rounded off to the nearest 0.1 Å.

[b] Functional group positions are identified to be consistent with Fig. 2.3 and are used to indicate chemically or spatially similar positions in the model substrate, model transition state, or inhibitors.

[c] For interactoins between blLAP residues and inhibitor residues which extend beyond the P1' position, see reference no. 30.

[d] For the LeuP structure there were 2 protomers per asymmetric unit. The interatomic distances shown are average values. The individual values were within 0.1 Å of each other in most cases.

[e] Values shown in parenthesis are generally not shown as interactions in the Figures due to the long internuclear distances.

[f] Ph = phosphonate, numbers of oxygens or waters are as in reference no.2

[g] wid = within interaction distance

[h] Distances derived from modeling analogous atoms of PheLeu or Leucyl phosphonate into the active site.

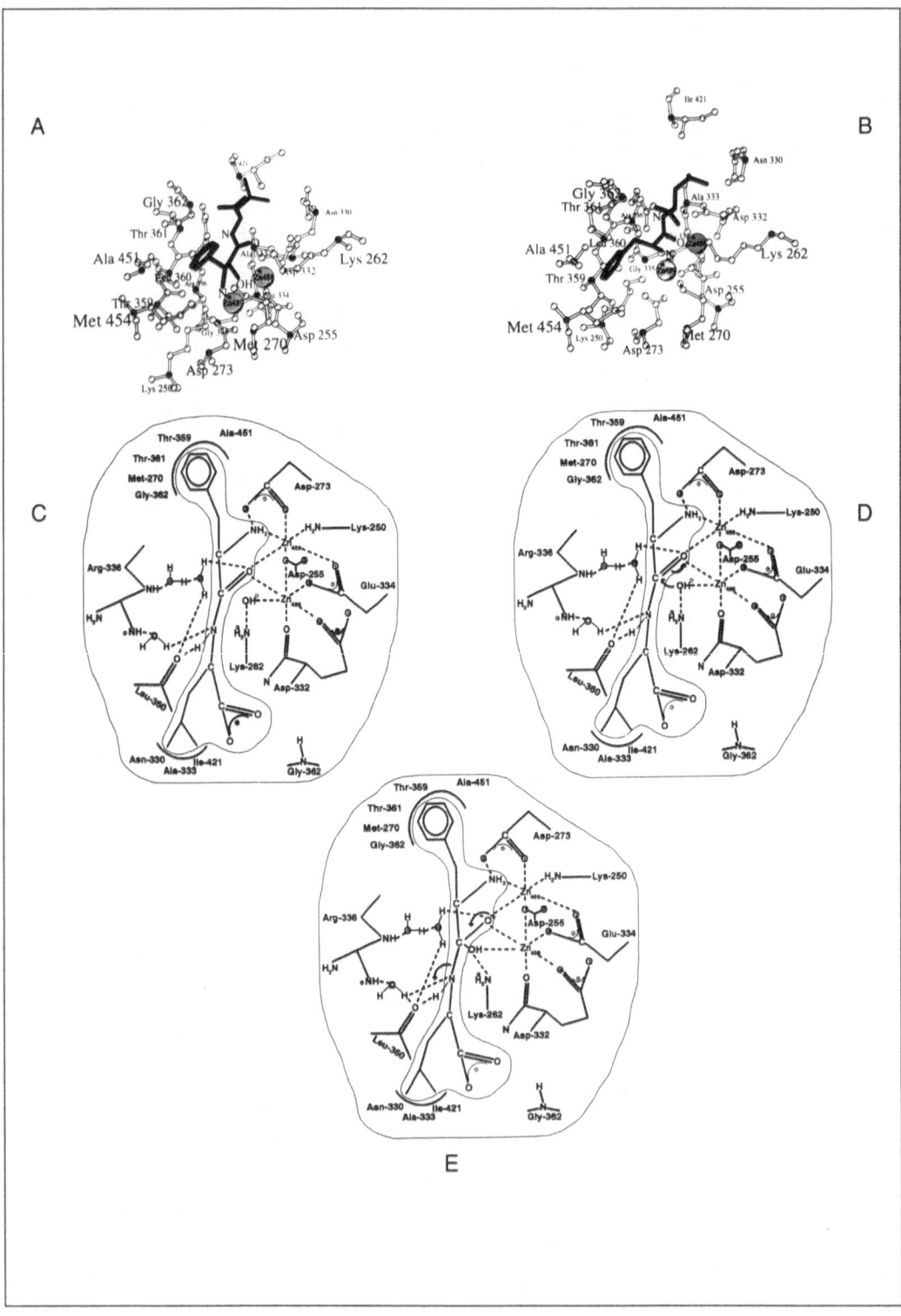

(Table 2.3). Within 3 Å of the P_1 residue only Thr-359, Asp-273, and Asp-255 are indicated. When only distances from the side chain are measured only Thr-359 is within 3 Å. The Met and both Thr were indicated as forming a hydrophobic binding pocket (presumably S_1, according to Schechter and Berger[47]) to which the phenylalanyl side chain of bestatin is bound in the blLAP-betatin complex.[10] For the bestatin blLAP complex, the following residues are within 4 Å of the P_1 residue of bestatin: *Lys-250, *Asp-255, Met-270, *Asp-273, Thr-359, Leu-360, Thr-361, Gly-362, Ala-451 and *Zn^{2+}489. For this complex only *Asp-273 is found within 3 Å. (The asterisks indicate contacts to the backbone.) Met-454 is also listed as part of the residues which comprise the P_1 binding pocket in the blLAP-bestatin complex, but this is not indicated for the blLAP-amastatin complex or any other complex. Recent searches of the data base confirm that Met-454 is >4 Å from bestatin. For the blLAP-LeuP complex the enzyme residues that are within 4 Å of the P_1 side chain are Thr-359, Asp-273, Met-270, Asp-332, Ala-451 and Lys-262. For the blLAP-leucinal complex Thr-359, Asp-273, Asp-255, Gly-362, Leu-360, Lys-250, Met-270, Glu-334, Asp-332, Ala-451 and Lys-262 are within 4 Å. Asp-273, Asp-255,

Fig. 2.4 (opposite). Binding of inhibitors and substrates to blLAP and proposed mechanism of hydrolysis of a model substrate, Phe-Leu.[2,9,39,40]

A. Mode of binding of bestatin to blLAP.[10]

B. Proposed mode of binding of the model substrate, PheLeu to blLAP. By analogy to the observed binding of bestatin, the side chains of the putative substrate PheLeu, are found within pockets formed by Met-270, Thr-359, Gly-362, Ala-451, and Met-454 (S_1), and by Ala-330, Ala-333, and Ile-421 (S_1'), respectively (see text). The carbonyl oxygen assumes a position similar to the position occupied by the hydroxyl in bestatin. Zn488 is the more readily exchanged ion. Both Zn ions appear to be involved in catalysis. In order to enhance clarity, PheLeu is shown in black. Zinc ions are in grey. Enzyme residues are in outline, and α-carbons of LAP amino acid residues are black. The blLAP PheLeu model is rotated slightly to permit visualization of the aromatic ring.

C. Polarization of the scissile carbonyl involves interactions with one or both Zn^{2+} and a hydrogen bond to the εNH₂ of Lys-262. Polarization of the scissile C-N bond and enhanced electrophilicity of the carbon of this bond involving interactions of the P_1' NH with the carbonyl oxygen of Leu-360 and the electrons of this amide with a proton from a water molecule (labeled C and found in the figure near Leu-360). Enzyme residues are against a grey background. PheLeu is shown against a white background. Labels for the water molecules are within the respective oxygen atom. Water C is between Leu-360 and Arg-336.

D. Formation of the transition state involves nucleophilic attack (arrows) at the scissile carbon by stabilized OH⁻ or H_2O (see text). The tetrahedral intermediate is stabilized by interaction of the negatively charged oxygen with either Zn^{2+}488 and the εNH₂ of Lys-262 or by both ions and water.

E. Hydrolysis is accomplished when the transition state collapses (arrows).

Glu-334, Asp-332 and Lys-262 are within 3 Å of the leucyl residue of the inhibitor, but as with the amastatin complex few, if any, of these would appear to comprise the hydrophobic binding pocket.

For binding the P_1',leucyl, side chain of bestatin, another putatively hydrophobic cleft (S_1') is indicated. It includes residues Asn-330, Ala-333 and Ile-421 (Fig. 2.4A). The latter two residues are 4 Å from the bestatin leucine side chain. Arg-336 and Asn-330 are 3.5 and 3.6 Å away, respectively. For binding of the amastatin P_1' residue, Val, enzyme residues Gly-362, Thr-361, Leu-360, Arg-336 and Asp-332 are within 4 Å. Only Gly-362 is within 3 Å. None of the enzyme residues which are listed as constituents of the S_1' pocket in the blLAP-bestatin complex seem to be within 4 Å of the P_1' residue of amastatin. Clearly, other residues are involved in the binding pockets, and longer (>4 Å) inter-residue distances (i.e., for Asn-330, Ala-333 and Ile-421) in the blLAP-amastatin complex are obtained. This suggests that the binding pockets are larger than most of the peptide side chains used.

This would appear to be consistent with the ability of LAP to hydrolyze aminoacyl naphthyl amides and large polypeptides.[21] The apparent hydrophobic nature of this binding pocket should be more thoroughly studied.

The backbone of the inhibitor is stabilized in the enzyme by hydrogen bonds involving Lys-262, Asp-273 and Leu-360.

As noted above, there is little shift in conformation of blLAP upon bestatin binding, and comparable binding of amastatin, LeuP, and leucinal are also indicated since overall alteration in structure of the protomers with different inhibitors bound is measured in tenths of an Å. For example the rms differences between the backbone atom positions of the native and bestatin complexed structures is 0.18 Å. This is comparable to the estimated precision of the refined atomic coordinates.[10] In addition, despite predictions to the contrary,[48] these αNH2-terminal D-residue-bearing inhibitors seem to bind in the same N-C orientation to blLAP as do analogous compounds bearing αNH$_2$-terminal L-residues.[20] This indicates that data obtained from these structural studies should be relevant in anticipating mechanisms of action of natural peptides, which can be expected to bind similarly.

Given the apparent structural stability of blLAP, several observations remain enigmatic. The markedly tighter binding of bestatin and amastatin than natural peptides with L-amino acid residues at their amino terminus might not have been anticipated, since in each of the "statins," the αNH2-residue is in the D-configuration, and LAP does not hydrolyze peptides or peptide analogs with αNH2-terminal residues which are of the D-configuration.[36-38] In addition there seems to either be insufficient precision in the structural data and/or enough flexibility in the enzyme such that the extra structure present in the "statin" inhibitors is easily modeled in the active site. Thus, there is sufficient overlap of the scissile peptide bond in substrates and the peptide bond in these inhibitors to allow mechanistic extrapolation (see mechanism of action section).

METAL IONS IN STRUCTURE AND CATALYSIS

Many, but not all, APs have been identified as zinc or cobalt metalloenzymes[40] (Table 2.1). Carpenter and Vahl (1973)[5] showed that native lens LAP protomers bind two zinc ions, each with different avidity. blLAP is active only when both of these metal-ion binding sites are occupied. Stoichiometric and kinetic data indicate that Mn^{2+}, Mg^{2+} and Co^{2+} can readily be exchanged for one zinc ion.[5,49] Structural data of blLAP in which one of the Zn^{2+} has been replaced by Mg^{2+} indicates that this Zn^{2+}, called the more readily exchangeable ion, is in position number 488.[31] By analogy it would appear that Mn^{2+} binds to this site as well. Since replacement of metals affected k_{cat} and was within a hydration radius of the substrate scissile carbonyl,[41] it was assumed that at least one of the metal ions was bound at the active site (see below).

For blLAP with two Zn^{2+} the apparent dissociation constant ratios, K_{Mg}/K_{Zn} at pH 9.5 and K_{Mn}/K_{Zn} at pH 8.5, are 150 and 35, respectively.[5] Co^{2+} can also replace Zn^{2+} at the more tightly bound site, number 489.[49] Replacement of Zn^{2+} with Co^{2+} at each site has different effects on K_m and k_{cat} for some substrates. LAP with Zn^{2+}488 and Co^{2+}489 has K_m and k_{cat} for leucineamide of 3.1 mM and 23 μmolmin^{-1}mg^{-1}, respectively, whereas the K_m and k_{cat} for LAP with Co^{2+}488, Zn^{2+}489 are 20 mM and 39 μmolmin^{-1} mg^{-1}.[49] (Table 2.4.) That substitution at both metal ion sites in blLAP affects both K_m and k_{cat}[49] (Table 2.4) suggested that both metal ions have a (possibly direct) role in binding and catalysis. Structural data support this possibility (Fig. 2.4). However, when leucyl-p-nitroanilide is used as substrate these constants are indistinguishable for the $Zn^{2+}Co^{2+}$ or $Co^{2+}Zn^{2+}$ forms of the enzyme (Table 2.4). These data

Table 2.4. K_m *and* k_{cat} *of LAP with different metal ion compositions*[49]

Enzyme Form		L-leucine-p-nitroanilide		L-leucine amide	
Site 1	Site 2	K_m (mM)	k_{cat} µmol min^{-1} mg^{-1}	K_m (mM)	k_{cat} µmol min^{-1} mg^{-1}
Zn^{2+}	Zn^{2+}	5.9 ± 0.6	0.13 ± 0.01	51 ± 8	43 ± 2
Mg^{2+}	Zn^{2+}	2.6 ± 0.5	2.8 ± 0.6	14 ± 3	400 ± 40
Co^{2+}	Co^{2+}	0.26 ± 0.03	0.26 ± 0.01	0.35 ± 0.05	59 ± 3
Mg^{2+}	Co^{2+}	0.90 ± 0.1	2.7 ± 0.2	2.0 ± 0.3	42 ± 5
Co^{2+}	Zn^{2+}	1.2 ± 0.2	0.20 ± 0.01	20 ± 3	39 ± 2
Zn^{2+}	Co^{2+}	1.1 ± 0.1	0.18 ± 0.01	3.1 ± 0.5	23 ± 3

indicate the subtlety of relationships between the metal ions in the enzyme and the mechanism of hydrolysis of different substrates. Similar conclusions can be made when other metal ions occupy the metal binding sites. These data also emphasize the need to consider subtleties of structure and composition in order to elucidate differences in rates of hydrolysis or binding constants when different substrates are used. A two Cd^{2+} enzyme which is inactive has also been observed.[49]

Based on crystallographic studies,[2,10] Zn^{2+}488 is shown liganded by residues Asp-255 and Asp-332 (one oxygen from a carboxylate and an oxygen from the backbone peptide bond), as well as by Glu-334 (Fig. 2.4, Table 2.5). Essentiality of Glu-334 is indicated by the observation that mutation of Glu-334 to Ala in a homologous *pepA-* encoded AP was associated with inactivation.[50] Another ligand of the metal ions in native enzyme might be a water. In the blLAP substrate complex, and in the putative transition state, the nucleophilic hydroxide bound at the tetrahedral carbon and the negatively charged oxygen derived from the scissile carbonyl could also be liganded to metal ions. Metal ion ligands in the native-, bestatin-, and presumptive PheLeu and Phe Leu-transition state are shown in Table 2.5.

Zn^{2+}489 is shown liganded by Asp-255, Lys-250, Asp-273 and Glu-334[2,5,10,22,31,39,40] (Fig. 2.4, Table 2.5). It is to Zn^{2+}489 that the αNH$_2$ group in inhibitors (Tables 2.3 and 2.5, Fig. 2.4), and presumably in substrates, is coordinated. A bridging water may also be coordinated to this ion in the native enzyme. In blLAP-inhibitor complexes this position would be occupied by one of the

Table 2.5. *Ligands to Zn²⁺488 and Zn²⁺489 in blLAP complexed with bestatin and with a hypothetical substrate PheLeu bound in the ground and the PheLeu transition states[a]*

Metal ion	Native	Bestatin	PheLeu[b]	PheLeu transit state
$Zn^{2+}488$	O_2 of Asp-332 COO	O_2 of Asp-332 COO	O_2 of Asp-332 COO	O_2 of Asp-332 COO
	Asp-332 C=O	Asp-332 C=O	Asp-332 C=O	Asp-332 C=O
	O_2 of Glu-334 COO	O_2 of Glu-334 COO	O_2 of Glu-334 COO	O_2 of Glu-334 COO
	O_1 of Asp-255 COO	O_1 of Asp-255 COO	O_1 of Asp-255 COO	O_1 of Asp-255 COO
		P_1 OH	O of P_1 C=O	gem diol OH
	H_2O #236	O of P_1 C=O or H_2O^c	activated OH⁻	gem diol OH or NH of P_1'
$Zn^{2+}489$	O_1 of Asp-273 COO	O_1 of Asp-273 COO	O_1 of Asp-273 COO	O_1 of Asp-273 COO
	εNH_2 of Lys-250	εNH_2 of Lys-250	εNH_2 of Lys-250	εNH_2 of Lys-250
	O_1 of Glu-334 COO	O_1 of Glu-334 COO	O_1 of Glu-334 COO	O_1 of Glu-334 COO
	O_1 of Asp-255 COO	O_1 of Asp-255 COO	O_1 of Asp-255 COO	O_1 of Asp-255 COO
		αNH_2 of P_1	αNH_2 of P_1	αNH_2 of P_1
	H_2O #236	OH of P_1	activated OH or O of P_1 C=O	OH of gem diolate

[a] This table differs from Kim and Lipscomb (1994)[30] in that no Zn^{2+}-Zn^{2+} bond is shown. Instead a water molecule in the ground state—and parts of the transition state—occupy the 6th position of each metal ion.[2]

[b] The interactions between the Zn^{2+} and ligands are hypothetical and derived from considerations of binding of bestatin, amastatin, LeuP and other inhibitors or from model building experiments using PheLeu or LeuVal as a hypothetical substrate.

[c] Assumed. Note that in the overlay of LeuP and bestatin there are 6 ligands for $Zn^{2+}488$ in the LeuP structure. Due to the long distance between the $Zn^{2+}488$ and the carbonyl oxygen in bestatin, these probably do not interact directly. Instead, a water is shown to replace that ligand, although it has not been demonstrated in crystallographic studies.

hydroxyls which is anticipated in the tetrahedral intermediate (Fig. 2.4).

Complex interactions that are partially similar between the two metal ions or between the ions and substrate are indicated for several APs. In blLAP, close proximity is indicated for the two metal ions in native enzyme and within enzyme-inhibitor complexes: 2.9 Å in native blLAP, and 3.1 Å, 3.3 Å and 3.5 Å in the blLAP-bestatin, blLAP-amastatin, and blLAP-LeuP complexes, respectively. Close proximity, "cocatalytic function", and an interaction of the two ions have been demonstrated for *Aeromonas* AP.[33,51] This is in keeping with the interactions described for blLAP.

In contrast with reports regarding equivalent or cocatalytic roles for both metals in *Aeromonas* AP and blLAP, are reports that only one ion appears to be required for activity, and addition of Mn^{2+} or Mg^{2+} does not affect the activities measured.[52] In *Aeromonas proteolytica*[51] the residues used to ligand the Zn^{2+} are Asp, Glu and His, and the metals are bridged over a 3.5 Å distance by Asp and H_2O.

In *E. coli* methionine AP, the two Co^{2+} are separated by 2.9 Å.[53] The ions are liganded by the side chains of Asp-97, Asp-108, Glu-204, Glu-235 and His-171 with approximately octahedral coordination.[53] In terms of both the novel backbone fold and the constitution of the active site, *E. coli* MAP appears to represent a new class of proteolytic enzymes.

As isolated, hkLAP appears to bind to a significant extent only one equivalent of zinc, but, like blLAP, it is activated upon additional binding of Mg^{2+} or Mn^{2+} ions and ions which do not bind to blLAP.[54] It is possible that the unoccupied site in the enzyme corresponds to that which occupies site 488 in blLAP. Additional similarity of hkLAP to blLAP is indicated, since Zn^{2+} occupancy is generally required at the site which is analogous to that occupied by $Zn^{2+}489$ in blLAP and since K_m and k_{cat} are affected by metal ion substitution.[49,55] It is interesting to speculate that of the 8% amino acid sequence difference between hkLAP and blLAP, there are differences in residues which alter metal binding.[13] Other than blLAP, there is no other example in which Lys is used to coordinate the metal ion (see section V).

A new observation is that of a third metal ion binding site in blLAP.[2] While unremarkable in a structural context, this is surprising since no prior compositional analyses indicated more than two Zn^{2+} per protomer.[5,42,49,56] This metal is coordinated by

Leu-170, Met-171, Thr-173, Arg-271, a water and possibly Met-274, and is 12 Å away from $Zn^{2+}489$. Any relation of this ion to the active site ions would presumably be via coordination to Arg-271 which is also coordinated to Asp-273. The latter is coordinated to $Zn^{2+}489$. It appears likely that the third metal ion, if it exists, is involved in stabilization of part of the interface between the NH_2-terminal and the catalytic domain in the protomer.[2]

Saccharomyces methionine AP also has zinc ions which are not at the active site.[57] However, while catalytic roles for these ions are not indicated, MAPs from which these Zn^{2+} finger binding sites have been removed are less effective in rescuing slow growth phenotype than is wild type MAP (see chapter 1).

MECHANISM OF ACTION

Ideally, a complete description of the mechanism should explain pH optimal for hydrolysis of esters, peptides, peptide analogs, large polypeptides, and why peptides are hydrolyzed at rapid rates despite considerable variability in their inhibition constants or K_m. It should also elucidate roles of both ions in K_m and k_{cat}, why some inhibitors bind in the usual fashion whereas others bind in slow and/or tight fashion, and why some substrates also act as activators of the hydrolysis of other substrates. The proposed mechanism of action should also explain how/why inhibitors with D-αNH_2-residues bind tightly to the enzyme; however, to be substrates for hydrolysis, an unblocked L-αNH_2 amino acyl residue is required.[38,58,59] Clearly not all of these details regarding the mechanism of action can be gleaned from static structural studies, particularly studies which employ tight-binding inhibitors. Nevertheless, substantial progress toward a mechanism of action which explains many of these phenomena has been achieved.

Knowledge that (1) each subunit can bind a bestatin or other substrate analog and (2) the enzyme structure does not change appreciably upon inhibitor binding, allowed use of kinetic, NMR, and x-ray diffraction information to infer a mechanism of hydrolysis of peptide substrates (Fig. 2.4B-E). The NMR data and crystallography suggest that the scissile carbonyl oxygen of substrates is coordinated to the more readily exchanged Zn^{2+} [2,31,39,41] (Fig. 2.4B,C). More recent data suggest that polarization of the carbonyl oxygen of the scissile peptide bond may also involve interaction with one or both Zn^{2+} ions and a hydrogen bond to the

εNH_2 of Lys-262[2,9,10,39-41] (Fig. 2.4C). Polarization of the scissile C-N bond and enhanced electrophilicity of the C of this bond involves interaction of the $P_1'NH$ with the carbonyl oxygen of Leu-360 and of the electrons of this amide with a proton from a water molecule (labeled C in Fig. 2.4). This water is coordinated to and probably activated by Arg-336. Accordingly, prior claims that Arg-336 is coordinated to the scissile carbonyl and that Arg-336 assumes different conformations in the native and inhibitor-complexed enzyme have been withdrawn.[2]

Hydrolysis involves nucleophilic attack at the carbon of the scissile carbonyl. There appear to be no enzyme nucleophiles in the area of the scissile peptide bond in blLAP, although this possibility has some support.[60] The absence of a nucleophile is consistent with an inability to inactivate or label the active site of blLAP using a variety of affinity labels, which required attack by an enzyme-bound nucleophile for covalent attachment.[22] Accordingly, general base catalysis was suggested for the mechanism of hydrolysis of peptides.[12] Recent data are also consistent with acid-assisted formation of the OH^- nucleophile.[2] The most plausible nucleophile would appear to be H_2O or stabilized OH^-. The nucleophilic OH^- is presumably generated by simultaneous dissociation of H_2O and stabilization of the incipient hydroxide ion by either both Zn^{2+} ions and a water, or by $Zn^{2+}488$ and εNH_2 of Lys-262. This origin of the OH^- complements prior proposals which implicated an unidentified enzyme-bound base for generation of the OH^-.[39] The tetrahedral intermediate formed upon hydroxide attack is stabilized by interaction of the negatively charged O with either $Zn^{2+}488$ and the εNH_2 of Lys-262 or by both Zn^{2+} ions and a water (Fig. 2.4D,E). This interaction may explain in part the effect of both ions on hydrolytic rates. The added OH^- remains stabilized by the interactions which aided its formation. Stabilization and departure of the leaving groups is accomplished by donation of a proton to the new NH_2 from a water (C) (Fig. 2.4C,E). Upon hydrolysis, an increase in pK of the αNH_2 to >9 would result in protonation and release from the active site. It is also plausible that the new leaving amine is protonated by Lys-262. For further considerations of these mechanisms, see reference 2. A previously suggested role for Asp-255 as a general base in the hydroytic process must be reevaluated.[31]

A two-step binding process was proposed[30] in order to rationalize the two K_is indicated for formation of the tight LAP-bestatin complex.[9] It is curious that a delay in blLAP-catalyzed hydrolysis is also observed using leucylamide as substrate (Taylor, unpublished). The two-step binding mechanism is also consistent with observations of two presteady-state intermediates in dimetal ion-containing hkLAP and with only one intermediate when hkLAP had one ion. It was proposed that the substrate would bind to both sites in dimetal forms of hkLAP but pass the unoccupied metal ion binding site and bind directly to the analog of $Zn^{2+}489$ in hkLAP to which only one equivalent of metal was bound.[30,54,55] For arginine AP the different binding kinetics and absence of metal content indicate other binding mechanisms exist.[45]

The mechanistic data presented above also allow rationalization of several unexpected observations. Thioamide derivatives did not show enhanced binding to hkLAP.[61] The LAP used for that work was in the $Zn^{2+}Mg^{2+}$ form. Assuming that hkLAP and blLAP have the same metal ion distribution, it would appear that Mg^{2+}, and not Zn^{2+}, was in the readily exchanged site. This might explain the unexpected decrease in apparent affinity found for thioamides in the hkLAP, since Mg^{2+} does not have the same binding affinity for sulfur-containing compounds as Zn^{2+}. This rationale would not appear to explain the lack of binding enhancement in the binding of thiobestatin and brain AP under conditions where zinc might be expected to be retained in the enzyme.[62] Nor would it be a sufficient rationalization of the data if both metals are equally involved in substrate binding. Another rationalization for the lack of effect may be because the SH of thiobestatin becomes oxidized and is, therefore, not complexed as tightly as the corresponding hydroxyl, with (one of) the metals. In addition, bestatin thioamide was also more weakly bound than bestatin to hkLAP.[61] This might be rationalized since, by analogy with bestatin, direct liganding of the sulfur with Zn^{2+} would not be expected. Alternate functions for the C2-SH or C2-OH in thiobestatin or bestatin, respectively, have been suggested.[61,62]

In contrast with the metallopeptidases, a critical active site serine is indicated in the D-amino acid AP of *Ochrobactrum anthropi* and some of the prolyl APs.[62a,b,c] The latter are presumably prolyl APs which are different from the prolyl APs which are very similar to, or identical to, metallopeptidases (see chapter 1 and section below).

HOMOLOGIES

Quantitative immunological techniques indicated that lens and kidney LAPs are indistinguishable within a particular organism.[13] This was corroborated since all of the amino acid sequence deduced for the bovine kidney enzyme was required to solve the blLAP structure.[1] Since those initial investigations, >100 references have been published in which homologies between APs are noted. Highlights from this literature are described here (consult chapters 3-8 for further information regarding homologies).

Immunological assays indicated that bovine and porcine LAP share 91% sequence homology, and bovine and human LAP share 81% sequence homology.[13] Furthermore, the shape of protomers and their arrangement within the hkLAP hexamer is indistinguishable from that observed in blLAP, although the two enzymes crystallize in different space groups.[8,14,63] Matsushima et al noted that the porcine intestinal prolyl-AP is indistinguishable from hkLAP and that human liver LAP is structurally very similar.[64] Rat kidney and brain prolyl-AP activity are also due to an enzyme which was indistinguishable from hkLAP.[65] Since the enzymes are also kinetically very similar, it is likely that they share common active site features (see below).

Many similarities between APs with respect to the use of two metals have been described above.[35,49,54] Monomeric microsomal AP-M (280 kDa) from porcine kidney also contains two Zn^{2+} which are involved in catalysis.[66,67] Monomeric AP-A (300 kDa) from porcine kidney sera is also present in other tissues of many species, is also a Zn^{2+} peptidase, and is similarly inhibited by amastatin[68,69] (Table 2.2).

Stirling et al noted sequence homologies between blLAP (487 amino acids) and the 55.3 kDa 503 amino acid *xerB* gene product and AP-A from *E. coli*. Also noted was comparable activation of the *E. coli* enzyme by Mn^{2+} and inhibition by EDTA (also see chapter 1).[5,70] Using the amino acid sequence provided by protein sequencing, Stirling et al[70] noted that *E. coli* AP-A has overall 31% identity of amino acids and 22% similar residues to blLAP. Identity of the protein is even greater (52%) in the C-terminal region. Using the deduced amino acid sequences,[1] we reported 18, 44, and 35% identities for the NH_2-terminal and C-terminal domains and for the entire protein, respectively.[10] For residue similarity, the numbers are 46, 66, and 60%, respectively. The higher degree of homology in the C-terminal region may be expected, since it is the C-terminal region which contains the active site of the bovine enzyme; thus, evolutionary constraints might be more stringent. All the residues which appear to be involved in zinc binding or catalysis in blLAP, the residues (except Met-454) used in inhibitor and presumably substrate-binding, organized secondary structure, and the loop regions comprising the active site are conserved between blLAP and *E. coli* pepA.

Identities between the *E. coli pepA* and *S. typhimurium pepA* genes[70] and between APs N and M (E.C. 3.4.11.2) were established using molecular genetic methods. P146 type II alveolar epithelial cell antigen is identical to AP-N.[71] The *xerB* gene product shows 49% identity with *Arabidopsis thaliana*.[72] Thus, there is homology of APs from plants to mammals.

Molecular genetic analysis indicates that AP-A (glutamyl aminopeptidase A) shares 92% and 86% sequence identity with the murine BP-1/6C3 and human gp160 differentiation antigens, respectively, with all the zinc binding residues being conserved.[73]

The data presented here suggest that blLAP, bkLAP, human lens and liver LAPs, hlLAP, hkLAP, porcine intestine LAP, prolyl-AP, *E. coli* AP-A, AP-I and the *S. typhimurium pepA* gene product are part of a new family of zinc APs which utilize the zinc-binding (and probably much of the substrate binding) amino acid constellations described for bovine LAP[1] (Fig. 2.2). Guenet et al report that blLAP and the two Zn^{2+} *Aeromonas* LAP share 18%-20% amino acid sequence identity, and that there is 45% similarity of the COOH-terminal domain of blLAP and the detergent-resistant alkaline exoprotease of *Vibrio alginolyticus*.[17] Like blLAP, *Aeromonas* LAP also binds bestatin and amastatin in a slow-, tight-fashion.[32,33] While primary sequence homology between blLAP and *Aeromonas* LAP is limited and amino acids which bind the metal ions are dissimilar, in both enzymes each subunit consists of two domains, with the active site located in the carboxyl terminal module. For blLAP and *Aeromonas* APs, the catalytic domain includes an eight-stranded β-sheet and seven structurally-equivalent helices. In both cases the two zinc ions form a binuclear cocatalytic center located at the end of strands 3 and 5 and helices E and G (Figs. 2.5 and 2.6). These peptidases can be distinguished from another recently identified superfamily of zinc proteases which appear to use Glu in catalysis, two His and Glu to bind zinc, and

```
BLLAP    TKGLVLGIYSKEKEEDEPQFTSAGENFNKLVSGKLREILNISGPPLKAGK    50
         |  |*|*||      |   ||| *|||*|| *  |  |* |     ** *
pepA     SACIVVGVF------EPRRLSPIAEQLDKISDGYISALLRRGELEGKPGQ    58
STRUC    SSSSS   DDD        HHHHHHHH    HHHHHH              SS


BLLAP    TRTFYGLHEDFPSVVVVGLGKKTAGIDEQENWHEGKENIRAAVAAGCRQI   100
         *   |   |     |   ||| |  *   |** |        |  |  *|
pepA     TLLLHHVPNVLSERILLIGCGKERELDERQYKQVIQKTINTLNDTGSMEA   108
STRUC    SSSSS       SSSSSSS       SS   SSHHHHHHHHHHHHHHHHHH


BLLAP    QDLEIPSVEVDPCGDAQAAAEGAVLGLYEYDDLKQKRKVVVSAKLHGSED   150
         |*|          *|* |** *   |*|  |** *   *||
pepA     VCFELHVKGRNNYWKVRQAVETAKETLYSFDQLKRKMVFNVPTRRELTSG   170
             LT                       TNKSEPRRPL
STRUC    HH      SSSSS    HHHHHHHHHH                  SSSSS


BLLAP    QEAWQRGVLFASGQNLARRLMETPANEMTPTKFAEIVEENLKSASIKTDV   200
         | *|*|*| | |*   *| * |   *|   |   | |  *     *| |
pepA     ERAIQHGLAIAAGIKAAKDLGNMPPNICNAAYLASQARQLADSYSKNVIT   220
STRUC    HHHHHHHHHHHHHHHHHHHHHHHH       HHHHHHHHHHHHHHHH  SSS


                                                        Z
BLLAP    FIRPKSWIEEQEMGSFLSVAKGSEEPPVFLEIHYKGSPNASEPPLVFVGK   250
         |       | * *|*|*| **|| || * *** **** *|*|***
pepA     RVIGEQQMKELGMHSYLAVGQGSQNESLMSVIEYKG--NASERPIVLVGK   270
STRUC    SSS HHHHH   HHHHHH    SSSSSSS        SSSSS


            Z     . B      B Z       .
BLLAP    GITFDSGGISIKAAANMDLMRADMGGAATICSAIVSAAKLDLPINIVGLA   300
         **|*********|  ** *| **|  ||  ||  |* |****|||*|
pepA     GLTFDSGGISIKPSEGMDEMKYDMCGAAAVYGVMRMVAELQLPINVIGVL   320
STRUC    SSSSS               HHHHHHHHHHHH       SSSSSS


                                   B ZBZ       .
BLLAP    PLCENMPSGKANKPGDVVRARNGKTIQVDWTDAEGRLILADALCYAHTFN   350
         ****|*|* |****|  |   * *||* **********|*|*|*|   *|
pepA     AGCENMPGGRAYRPGDVLTTMSGQTVEVLNTDAEGRLVLCDVLTYVERFE   370
STRUC    SSSSS        SSSSSS SSSSS      HHHHHHHHHHHHHH


            BB B        .
BLLAP    PKVIINAATLTGAMDIALGSGATGVFTNSSWLWNKLFEASIETGDRVWRM   400
         *  ||*||******  ****   **||| *     *  ** ||***|**|
pepA     PEAVIDVATLTGACVIALGHHITGLMANHNPLAHELIAASEQSGDRAWRL   420
STRUC    SSSSS       HHHHH       SSSSS  HHHHHHHHHHHH    SSS


                        .B
BLLAP    PLFEHYTRQVIDCQLADVNNIGKYRSAGACTAAAFLKEFVTHPKWAHLDI   450
         ** | *  * |||||**| ***   |** **||** *|   ******
pepA     PLGDEYQEQ-LESNFADMANIGG-RPGGAITAGCFLSRFTRKYNWAHLDI   469
STRUC    HHHHH       SSSS      HHHHHHHH          SSSSS


         B  B        .
BLLAP    AGVMTNKDEVPYLRKGMAGRPTRTLIEFLFRFSQDSA               487
         **|  |***| |**| |*||**| |  |
pepA     AGTAWRSGKA----KGATGRPVALLAQFLLNRAGFNGEE              503
STRUC    SS        SS    HHHHHHHHHHH DDD
```

Fig. 2.5. Comparison of the amino acid sequences of bovine lens and E. coli aminopeptidase A (pepA) showing the secondary structure of blLAP. (*) identity, (|) similarity; structure codes: S, β-strand; H, α-helix; D, disordered loop or terminus. Z indicates zinc ion binding residues. B indicates residues involved in binding bestatin. Adapted from Burley et al.[10]

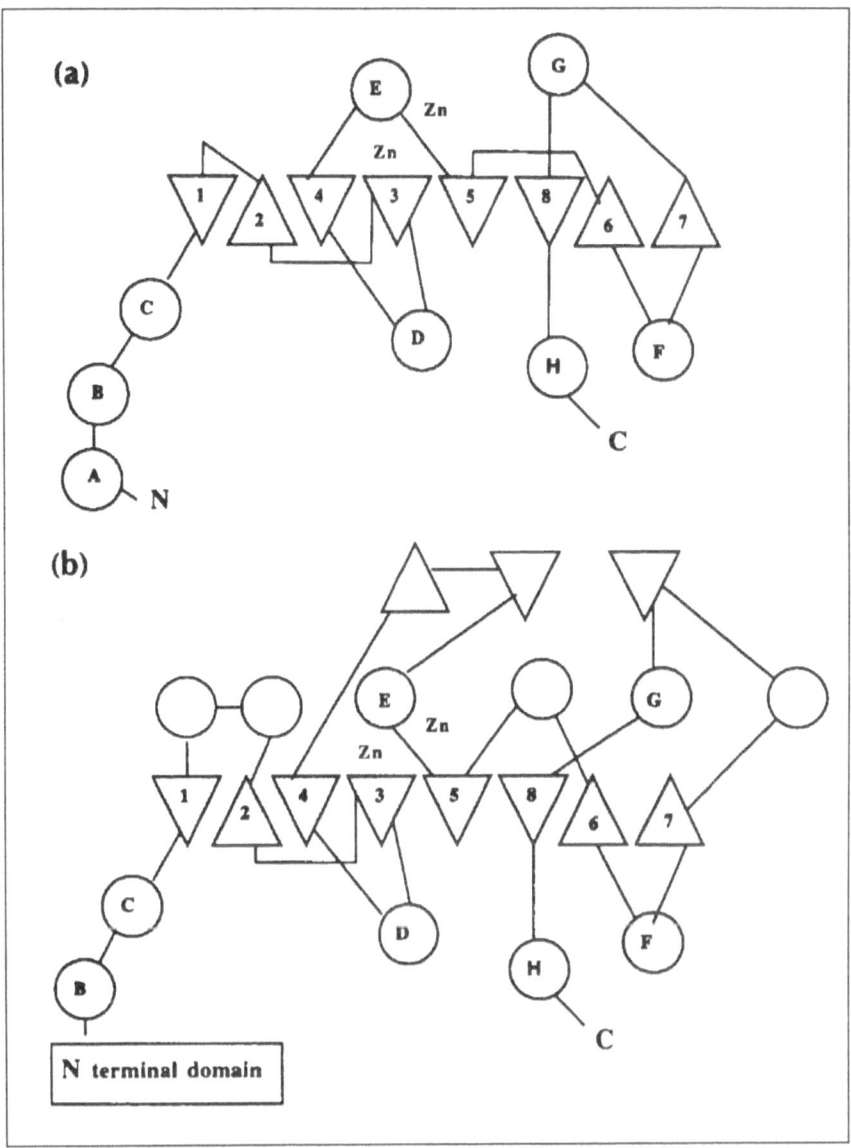

Fig. 2.6. Schematics comparing the topology of (a) Aeromonas proteolytica aminopeptidase and (b) the carboxyl terminal domain of blLAP. Circles represent α-helices and triangles represent β-strands. Elements with numbers or letters indicate structurally equivalent secondary structure elements. Adapted from Chevrier et al.[51]

Arg in substrate binding.[15] These include rat kidney zinc protease (rKZP), AP-N, thermolysin *B.T.* (*B. thermoproteolyticus*), thermolysin *B.S.* (*B. stearothermophilus*), protease *B.A.* (*B. amyloliquefaciens*), protease *Serratia*, rat enkephallinase, carboxypeptidases A and B and, possibly APs M and N, some collagenases, angiotensin-converting enzyme, human AP-M and leukotriene A4 hydrolase. The extent of similarity between rKZP and the last two enzymes is 77% and 31%, respectively. The extent of homology between rKZP and the *E. coli* AP-N is 18%.[74]

The deduced amino acid sequence from the *Saccharomyces* methionine AP shows approximately 40% sequence homology with methionine AP from *E. coli, S. Typhimurium, and B. subtilis.*[18] The yeast enzyme consists of two functional domains: a unique NH$_2$-terminal domain containing two motifs resembling zinc fingers, which may allow the protein to interact with ribosomes and function cotranslationally (also see refs. 75, 76), and a catalytic COOH-terminal domain resembling other procaryotic methionine APs. Most of the similarity between these enzymes is localized to the COOH region. These enzymes share little homology with blLAP[76] and, like AP-A, some are not inhibited by bestatin.

Streptococcus thermophilus CNR 302 contains at least three aminopetidases. One gene codes for a 445 amino acid residue, 50.4 kDa protein (total mass of 300 kDa; thus it is a hexamer) which is 70% identical to PepC from *Lactoccocus lactis cremoris* and shows 38% identity to eucaryotic bleomycin hydrolase.[77] It also contains regions of strong similarity to cysteine proteinases and appears to be a thiol AP. The *Lactobacillus helveticus pepC* gene product is encoded in two open reading frames, is 51.4 kDa, and shares 48 and 98% sequence identity with the PepC proteins from *Lactococcus lactis* and *L. helviticus* CNRZ32, respectively.[78]

Despite many common features between methionine AP, *Aeromonas* AP, and leucine AP there are sufficient differences to conclude that these enzymes probably did not evolve from a common ancestral protein, and they are representative of different classes of enzymes.

EXPRESSION

The advent of molecular genetic techniques resulted in a bourgeoning in reports regarding expression of aminopeptidases. Only some of that literature, primarily that which focuses on LAP, can be described here. Analysis of lens and kidney cells and tissue showed different patterns of expression of LAP mRNAs. Lens epithelial tissue showed only one transcript (2.4 kb), whereas two transcripts (2.0 and 2.4 kb) were observed in cultured lens cells and kidney tissue.[1] Since Southern blot analysis indicates only one gene for bovine LAP, our observation of two transcripts suggests that they arise by differential splicing of a common precursor RNA and that they code for similar but distinct proteins.[1] The C-terminus in the 2.0 kb transcript is distinct from that for the 2.4 kb transcript. As yet no smaller protein has been detected in kidney or lens by standard Western analysis or immunoprecipitations. It remains possible that both messages code for proteins of similar size which cannot be distinguished by SDS-PAGE. Since these hybridization results indicate that the amino acid sequences are different beyond residues ≈300, it would appear that Asp-332, Glu-334, Arg-336, Leu-360 and Thr-359, which are implicated in catalytic function and zinc or substrate binding in blLAP, are not present in the protein encoded by the 2.0 kb transcript in bkLAP and in blLAP in lens epithelial cells in culture. It is interesting that two AP activities have been identified in human liver,[79] and two different mRNAs, both originating from one gene, were also observed for kidney AP-N.[15]

Northern blot analysis indicates that LAP mRNA levels correlate with LAP activity in progressively passaged cells.[1] In this in vitro model of aging, there was an increase in message and activity at low passage, and this correlates with similar increased levels in intracellular proteolysis.[80] However, the two messages were not coordinately expressed. Whereas at early and later passages (passages 1, 2 and 5-14, respectively), there were roughly equal levels of the 2.0 and 2.4 kDa messages, the level of 2.0 message was ≈5-fold greater than levels of the 2.4 kDa message during passages 3 and 4.[80] This suggests that LAP or LAP-like protein plays a critical role in the intracellular proteolytic response to events occurring during early passages in lens epithelial cells in culture.[80] The data also indicate that the regulation of LAP expression is at least in part at the transcriptional level.

It is interesting to note that LAP activity and expression levels are also enhanced due to treatment with interferon γ in cells from human HS153 fibroblasts, ACHN renal carcinoma, A549 lung carcinoma, and A375 melanoma[81] (Fig. 2.7). Since the response is saturable, it appears to be receptor-mediated.

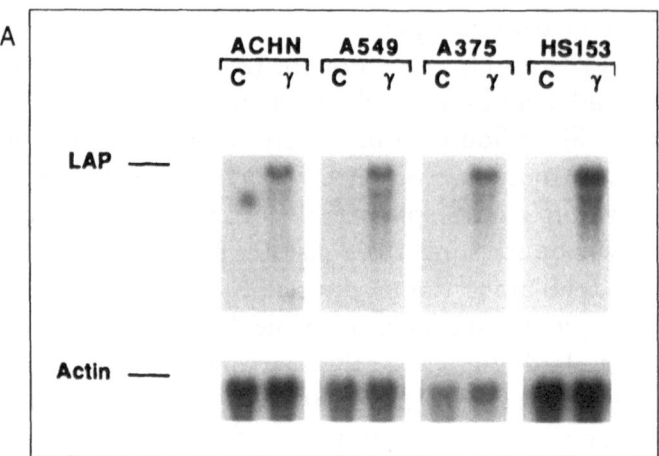

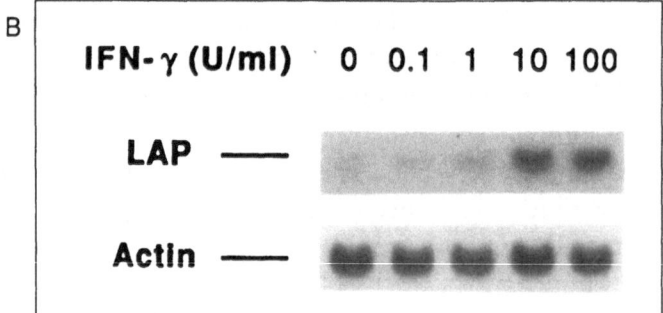

Fig. 2.7. (A) LAP mRNA is induced by IFN-γ in multiple cell types.
ACHN renal carcinoma cells, A549 human lung carcinoma cells,
A375 human melanoma cells, and HS153 human fibroblasts were
treated with control medium or with IFN-γ. RNAs separated on an
agarose gel and transferred to a nylon membrane were detected
by hybridization with LAP or β-actin cDNA probes. (B) Dose
response of induction of LAP mRNA by IFN-γ. ACHN human renal
carcinoma cells were treated with control medium or the indicated concentrations of IFN-γ.

Aging and differentiation are associated with variations in AP
expression. Equivalent levels of 2.4 kb LAP mRNA were found in
lens epithelia (the youngest part of the lens) from young and old
animals.[80] This is consistent with continued growth and protein
synthesis in this portion of the lens through life. No LAP mRNA
was observed in the cortex (older, more differentiated tissue than

epithelium). This is in keeping with the cortex being terminally differentiated, devoid of nuclei, and synthetically quiescent. It is in contrast with our observations of significant levels of LAP activity in this part of the lens, particularly in lenses from young animals.[82]

REMAINING MYSTERIES AND CONCLUSIONS

The aggregate of compositional, structural and kinetic data allows rationalization of many kinetic/binding data and provides significant insight into the mechanism of action of LAP. However, intriguing questions remain. Perhaps the most obvious questions are (1) what are roles for both metals and (2) what are the determinants of substrate specificity. The available proposals still fall short of a complete explanation of relationships between both metals, K_m and k_{cat}.

A clear explanation of the pH profile remains elusive. The greater activity of blLAP at pH >8 is consistent with the greater facility of replacing metal ions at elevated pH and with an unprotonated state of the substrate αNH_2.[5] It is also consistent with determinations of pH dependencies of inhibition and kinetic parameters for L-leucyl-p-nitroanilide and L-leucinal.[60] But it does not explain the broad pH range (pH 6-11) of activity of the enzyme.

Complete rationalization of the slow and tight binding of the "statin" type or other inhibitors is also not available. It is clear that the C2-OH in bestatin is involved in tighter binding than is observed in the analogous substrate peptide. A two-step binding mechanism is proposed for the slow and tight binding of bestatin to LAP. However, specific enzyme-inhibitor interactions for binding substrates and inhibitors remain to be established. It would be interesting to know if "slow" events for the inhibitors and substrate involve similar interactions in the enzyme, i.e., whether there is a two-step binding process involving formation of initial and tight complexes for each.[9]

Several compounds, some of which are substrates, were anticipated to be inhibitors of LAP. Surprisingly, several of these compounds are activators of blLAP or related enzymes. These include Leu- and Ala-p-aminobenzenesulfonates (which are also substrates), orthanilic and aminobenzenesulfonic acid,[23] and an amastatin analog [(3S,4S)-statin-Val-Val-Asp].[32] The observed activation was interpreted as evidence for a second, perhaps adjacent, binding site, but this remains to be established.[83]

Higher-resolution, structural information, perhaps with dynamic aspects, is needed to pursue these questions and to elucidate differences in design between enzymes specific for D-, as opposed to L-amino acids, and the questions noted above. The myriad of proven and putative functions of APs make a compelling reason to elucidate their structures and molecular mechanisms of action. In turn, the availability of structural and kinetic data will foster further appreciation of roles of APs in physiology.

ACKNOWLEDGMENTS

This project has been funded in part with federal funds from the United States Department of Agriculture under contract number 53-3K06-0-1; National Institutes of Health, grant number EYO8566; The Daniel and Florence Guggenheim Foundation; and Research Corporation. The help of E. Epstein in preparation of the manuscript and Dr. J. Jahngen-Hodge for editorial assistance is greatly appreciated. The author is indebted to Dr. W. Bachovchin for help with the computer graphics and modeling and to Drs. N. Strater and W. N. Lipscomb for sharing unpublished data.

REFERENCES
1. Wallner BP, Hession C, Tizard R et al. Isolation of bovine kidney leucine aminopeptidase cDNA: Comparison with the lens enzyme and tissue-specific expression of two mRNAs. Biochemistry 1993; 32:9296-301.
2. Strater N, Lipscomb WN. Transition state analogue L-leucine-phosphonic acid bound to bovine lens leucine aminopeptidase: X-ray structure at 1.65 Å resolution in a new crystal form. Biochemistry 1995; 34:9200-10.
3. Carpenter FH, Harrington KT. Intermolecular cross-linking of monomeric proteins and cross-linking of oligomeric proteins as a probe of quaternary structure. J Biol Chem 1972; 247:5580-6.
4. Melbye SW, Carpenter FH. Leucine aminopeptidase (bovine lens) stability and size of subunits. J Biol Chem 1971; 246:2459-63.
5. Carpenter FH, Vahl JM. Leucine aminopeptidase (bovine lens): Mechanism of activation by Mg^{2+} and Mn^{2+} of the zinc metalloenzyme, amino acid composition, and sulfhydryl content. J Biol Chem 1973; 248:294-304.
6. Cuypers HT, Van Loon-Klaassen LAH, Vree Egberts WTM et al. The primary structure of leucine aminopeptidase from bovine eye lens. J Biol Chem 1982; 257:7077-85.
7. Taylor A, Carpenter FH, Wlodawer A. Leucine aminopeptidase (bovine lens): an electron microscopic study. J Ultrastruct Res 1979;

68:92-100.

8. Jurnak F, Rich A, van Loon-Klaassen L et al. Preliminary x-ray study of leucine aminopeptidase (bovine lens): an oligomeric metalloenzyme. J Mol Biol 1977; 112:149-53.

9. Taylor A, Peltier CZ, Torre FJ et al. Inhibition of bovine lens leucine aminopeptidases by bestatin: number of binding sites and slow binding of this inhibitor. Biochemistry 1993; 32:784-90.

10. Burley SK, David PR, Sweet RM et al. Structure determination and refinement of bovine lens leucine aminopeptidase and its complex with bestatin. J Mol Biol 1992; 224:113-40.

11. Chernaya MM, Nurbekov MK, Fedorov AA. Study of leucine aminopeptidase crystals from bovine pancreas. Biofizika 1985; 30:700-1.

12. Nishizawa R, Saino T, Takita T. Synthesis and structure-activity relationships of bestatin analogues, inhibitors of aminopeptidase. B J Med Chem 1977; 20:510-15.

13. Taylor A, Surgenor T, Thomson DKR. Comparison of (concentration and amino acid sequence homology between) leucine aminopeptidase in human lens, beef lens, beef kidney, hog lens, and hog kidney. Exp Eye Res 1984; 38:217-29.

14. Taylor A, Volz K, Lipscomb WN et al. Leucine aminopeptidase from bovine lens and hog kidney: Comparison using immunological techniques, electron microscopy and x-ray diffraction. J Biol Chem 1984; 259:14757-61.

15. Watt VM, Yip CC. Amino acid sequence deduced from a rat kidney cDNA suggests it encodes the Zn^{2+}-peptidase aminopeptidase N. J Biol Chem 1989; 264:5480-7.

16. Cueva R, Garcia-Alvarez N, Suarez-Rendueles P. Yeast vacuolar aminopeptidase yscI isolation and regulation of the *APE1 (LAP4)* structural gene. FEBS Lett 1989; 259:125-9.

17. Guenet C, Lepage P, Harris BA. Isolation of the leucine aminopeptidase gene from *Aeromonas proteolytica*. J Biol Chem 1992; 267:8390-5.

18. Chang YH, Teichert U, Smith JA. Molecular cloning, sequencing, deletion, and overexpression of a methionine aminopeptidase gene from *Saccharomyces cerevisiae*. J Biol Chem 1992; 267:8007-11.

19. van Loon-Klaassen L, Cuypers HT, Bloemendal H. Limited tryptic digestion of bovine eye lens leucine aminopeptidase. FEBS Lett 1979; 107:366-70.

20. Kim H, Lipscomb WN. X-ray crystallographic determination of the structure of bovine lens leucine aminopeptidase complexed with amastatin: formulation of a catalytic mechanism featuring a *gem*-diolate transition state. Biochemistry 1993; 32:8465-78.

21. Taylor A, Daims MA, Lee J et al. Identification and quantification of inactive leucine aminopeptidase in aged normal and cataractous human lenses and ability of bovine lens LAP to cleave bovine crystallins. Curr Eye Res 1982; 2:47-56.

21a. Niven GW. Purification and characterization of aminopeptidase A from *Lactococcus lactis* subsp. *lactis* NCDO712. J Gen Microbiol 1991; 137:1207-12.

22. Taylor A, Peltier CZ, Jahngen EGE Jr et al. Use of azidobestatin as a photoaffinity label to identify the active site peptide of leucine aminopeptidase. Biochemistry 1992; 31:4141-50.

23. Taylor A, Tisdell FE, Carpenter FH. Leucine aminopeptidase (bovine lens): synthesis and kinetic properties of ortho, meta, and para substituted leucyl-anilides. Arch Biochem Biophys 1981; 210:90-7.

24. Botbol V, Scornik OA. Measurement of instant rates of protein degradation in the livers of intact mice by the accumulation of bestatin-induced peptides. J Biol Chem 1991; 266:2151-7.

25. Squire CR, Talebian M, Menon JG et al. Leucine aminopeptidase-like activity in *Aplysia hemolymph* rapidly degrades biologically active alpha-bag cell peptide fragments. J Biol Chem 1991; 266:22355-63.

26. Cha S. Tight-binding inhibitors - III. A new approach for the determination of competition between tight-binding inhibitors and substrates—inhibition of adenosine deaminase by conformycin. Biochem Pharmacol 1976; 25:2695-702.

27. Morrison JF, Walsh CT. The behavior and significance of slow-binding enzyme inhibitors. [Review]. Advances in Enzymol Related Areas of Mol Biol 1988; 61:201-301.

28. Williams JW, Morrison JF. The kinetics of reversible tight-binding inhibition. Methods in Enzymol 1979; 63:437-67.

29. Patterson EK. Inhibition by bestatin of a mouse ascites tumor dipeptidase. J Biol Chem 1989; 264:8004-11.

30. Kim H, Lipscomb WN. Structure and mechanism of bovine lens leucine aminopeptidase. [Review] Adv Enzymol & Related Areas. Mol Biol 1994; 68:153-213.

31. Kim H, Lipscomb WN. Differentiation and identification of the two catalytic metal binding sites in bovine lens leucine aminopeptidase by x-ray crystallography. Proc Natl Acad Sci USA 1993; 90:5006-10.

32. Rich DH, Moon BJ, Harbeson S. Inhibition of aminopeptidases by amastatin and bestatin derivatives. Effect of inhibitor structure on slow-binding processes. J Med Chem 1984; 27:417-22.

33. Wilkes SH, Prescott JM. The slow, tight binding of bestatin and amastatin to aminopeptidases. J Biol Chem 1985; 260:134154-62.

34. Peltier CZ, Taylor A. Kinetic parameters for the slow binding of bestatin to beef lens leucine aminopeptidase (bLAP). Fed Proc 1986; 45:1856 (abstr).

35. Rohm KH. Interaction of the (2S,3S)-isomer of bestatin with yeast aminopeptidase I. Kinetic and binding studies. Hoppe-Seylers Zeitschrift fur Physiologische Chemie 1984; 365:1235-46.

36. Smith EL, Hill RL. Leucine aminopeptidase. In: Boyer PD, ed.

The Enzymes. 2nd ed, vol 4 (part A). New York: Academic Press, 1960:37-63.

37. Delange RJ, Smith EL. Leucine aminopeptidase and other N-terminal exopeptidases. In: Boyer PD, ed. The Enzymes. Vol. 3. New York: Academic Press, 1971:81-103.

38. Hanson JH, Frohne M. Crystalline leucine aminopeptidase from lens. Methods Enzymol 1976; 45:504-10.

39. Taylor A. Aminopeptidases: toward a mechanism of action. TIBS 1993; 18:167-172.

40. Taylor A. Aminopeptidases: structure and function. FASEB J 1993; 7:290-298.

41. Taylor A, Sawan S, James T. On the binding of leucyl-o-sulfonic acid in leucine aminopeptidase. Interaction between this substrate analog and the activation site metal-viewed by NMR. J Biol Chem 1982; 257:11571-6.

42. Thompson GA, Carpenter FH. Leucine aminopeptidase (bovine lens) the relative binding of cobalt and zinc to leucine aminopeptidase and the effect of cobalt substitution on specific activity. J Biol Chem 1976; 251:1618-24.

43. Bryce GF, Rabin BR. The function of the metal ion in leucine aminopeptidase and the mechanism of action of the enzyme. Biochem J 1964; 90:513-8.

44. Nishino N, Powers JC. Design of potent reversible inhibitors for thermolysin. Peptides containing zinc coordinating ligands and their use in affinity chromatography. Biochemistry 1979; 18:4340-7.

45. Harbeson SL, Rich DH. Inhibition of arginine aminopeptidase by bestatin and arphamenine analogues. Evidence for a new mode of binding to aminopeptidases. Biochemistry 1988; 27:7301-10.

46. Ricci JS Jr, Bousvaros A, Taylor A. Crystal and molecular structure of bestatin and its implications regarding substrate binding to the active site of leucine aminopeptidase. J Org Chem 1982; 47:3063-5.

47. Schechter I, Berger A. Characterization of a pyroglutamate aminopeptidase from rat serum that degrades thyrotropin-releasing hormone. Biochem Biophys Res Commun 1967; 27:157-62.

48. David PR. Ph.D. thesis, Harvard University, Cambridge, MA, 1991.

49. Allen MP, Yamada AH, Carpenter FH. Kinetic parameters of metal-substituted leucine aminopeptidase from bovine lens. Biochemistry 1983; 22:3778-83.

50. McCulloch R, Burke ME, Sherratt DJ. Peptidase activity of *Escherichia coli* aminopeptidase A is not required for its role in Xer site-specific recombination. Mol Microbiol 1994; 12:241-51.

51. Chevrier B, Schalk C, D'Orchymont H et al. Crystal structure of *Aeromonas proteolytica* aminopeptidase: a prototypical member of the co-catalytic zinc enzyme family. Structure 2 1994; 283-91.

52. Prescott JH, Wagner FW, Holmquist B. Spectral and kinetic stud-

ies of metal-substituted *Aeromonas* aminopeptidase: nonidentical, interacting metal-binding sites. Biochemistry 1985; 24:5350-6.

53. Roderick SL, Matthews BW. Structure of the cobalt-dependent methionine aminopeptidase from *Escherichia coli*: a new type of proteolytic enzyme. Biochemistry 1993; 32:3907-12.

54. Van Wart HE, Lin SH. Metal binding stoichiometry and mechanism of metal ion modulation of the activity of porcine kidney leucine aminopeptidase. Biochemistry 1981; 20:5682-9.

55. Lin W-Y, Lin SH, Morris R et al. Stopped-flow cryoenzymological investigation of the presteady-state kinetics of hydrolysis of leu-gly-NHNH-Dns by leucine aminopeptidase. Biochemistry 1988; 27:5068-74.

56. Thompson GA, Carpenter FH. Leucine aminopeptidase (bovine lens) the relative binding of cobalt and zinc to leucine aminopeptidase and the effect of cobalt substitution on specific activity. J Biol Chem 1976; 251:1618-24.

57. Zuo S, Guo Q, Ling C et al. Evidence that two zinc fingers in the methionine aminopeptidase from *Saccharomyces cerevisiae* are important for normal growth. Mol Gen Genet 1995; 246:247-53.

58. McDonald JK, Barrett AJ. In: McDonald JK, Barrett AJ, eds. Mammalian Proteases. Vol 2. New York: Academic Press, 1986.

59. Smith EL, Spackman DH. Leucine aminopeptidase V. Activation, specificity, and mechanism of action. J Biol Chem 1995; 271-99.

60. Andersson L, Isley TC, Wolfenden R. α-aminoaldehydes: Transition state analogue inhibitors of leucine aminopeptidase. Biochemistry 1982; 21:4177-80.

61. Ocain TD, Rich DH. Synthesis of sulfur-containing analogues of bestatin. Inhibition of aminopeptidases by alpha-thiolbestatin analogues. J Med Chem 1988; 31:2193-9.

62. Gordon EM, Godfrey JD, Delaney NG et al. Design of novel inhibitors of aminopeptidases. Synthesis of peptide-derived diamino thiols and sulfur replacement of analogues of bestatin. Biochemistry 1988; 31:2199-2211.

62a. Gonzales T, Robert-Baudouy J. Bacterial aminopeptidases: properties and functions. FEMS Microbiol Rev 1996; in press.

62b. Asano Y, Nakazawa A, Kato Y et al. Properties of a novel D-stereospecific aminopeptidase from *Ochrobactrum anthropi*. J Biol Chem 1989; 264:14233-39.

62c. Asano Y, Kato Y, Yamada A et al. Structural similarity of D-aminopeptidase to carboxypeptidase DD and β-lactamases. Biochemistry 1992; 31:2316-28.

63. Taylor A, Carpenter FH, Wlodawer A. Leucine aminopeptidase (bovine lens): an electron microscopic study. J Ultrastruct Res 1979; 68:92-100.

64. Matsushima M, Takahashi T, Ichinose M et al. Purification and characterization of prolyl aminopeptidases from pig intestinal mu-

cosa and human liver. Structural, immunological and enzymatic evidence for identity with leucyl aminopeptidase. Biochem Biophys Res Commun 1991; 178:1459-64.

65. Turzynski A, Mentlein R. Prolyl aminopeptidase from rat brain and kidney. Eur J Biochem 1990; 190:509-15.

66. Pfleiderer G. Particle-bound aminopeptidase from pig kidney. Methods Enzymol 1970; 19:514-21.

67. Wacker H, Lehky P, Fischer EH et al. Physical and chemical characterization of pig kidney particulate aminopeptidase. Helv Chim Acta 1971; 54:473-85.

68. Tobe H, Kojima F, Aoyagi T et al. Purification by affinity chromatography using amastatin and properties of aminopeptidase A from pig kidney. Biochim Biophys Acta 1980; 613:459-68.

69. Wang J, Cooper MD. Histidine residue in the zinc-binding motif of aminopeptidase A is critical for enzymatic activity. Proc Natl Acad Sci USA 1993; 90:1222-6.

70. Stirling VJ, Colloms SD, Collins JF et al. *xerB*, an *Escherichia coli* gene required for plasmid ColE1 site-specific recombination, is identical to *pepA*, encoding aminopeptidase A, a protein with substantial similarity to bovine lens leucine aminopeptidase. EMBO J 1989; 8:1623-7.

71. Funkhouser JD, Tangada SD, Jones M et al. Type II alveolar epithelial cell antigen is identical to aminopeptidase N. Am J Physiol 1991; 260:L274-9.

72. Bartling D, Weiler EW. Leucine aminopeptidase from *Arabidopsis thaliana*. Molecular evidence for a phylogenetically conserved enzyme of protein turnover in higher plants. Eur J Biochem 1992; 205:425-31.

73. Song L, Ye M, Troyanovskaya M et al. Rat kidney glutamyl aminopeptidase (aminopeptidase A): molecular identity and cellular localization. Am J Physiol 1995; 267:F546-57.

74. Malfroy B, Kado-Fong H, Gros C et al. Molecular cloning and amino acid sequence of rat kidney aminopeptidase M: a member of a super family of zinc-metallohydrolases. Biochem Biophys Res Commun 1989; 161:236-41.

75. Arfin SA, Bradshaw RB. Cotranslational processing and protein turnover in eucaryotic cells. Biochemistry 1988; 27: 7979-84.

76. Ben-Bassat A, Bauer K, Sheng-Yung C et al. Processing of the initiation methionine from proteins: properties of the *Escherichia coli* methionine aminopeptidase and its gene structure. J Bacteriol 1987; 169:751-7.

77. Chapot-Chartier MP, Rul F, Nardi M et al. Gene cloning and characterization of PepC, a cysteine aminopeptidase from *Streptococcus thermophilus*, with sequence similarity to the eucaryotic bleomycin hydrolase. Eur J Biochem 1994; 224:497-506.

78. Vesanto E, Varmanen P, Steele JL et al. Characterization and ex-

pression of the *Lactobacillus helveticus* pepC gene encoding a general aminopeptidase. Eur J Biochem 1994; 224: 991-7.

79. Ledeme N, Vincent-Fiquet O, Hennon G et al. Human liver l-leucine aminopeptidase: Evidence for 2 forms compared to pig liver enzyme. Biochimie 1983; 65:397-404.

80. Eisenhauer DA, Berger JJ, Peltier CZ et al. Protease activities in cultured beef lens epithelial cells peak and then decline upon progressive passage. Exp Eye Res 1988; 46: 579-90.

81. Harris CA, Hunte-McDonough B, Krauss MR et al. Induction of leucine aminopeptidase by interferon gamma identification by protein microsequencing after purification by preparative two-dimensional gel electrophoresis. J Biol Chem 1992; 267: 6865-9.

82. Taylor A, Brown MJ, Daims MA, Cohen J. Localization of leucine aminopeptidase in hog lenses using immunofluorescence and activity assays. Invest Ophthalmol Vis Sci 1983; 24:1172-81.

83. DiGregorio M, Pickering DS, Chan WW-C. Multiple sites and synergism in the binding of inhibitors to microsomal aminopeptidase. Biochemistry 1998; 27:3613-7.

84. Woolford CA, Daniels LB, Park FJ et al. The *pep4* gene encodes an aspartyl protease implicated in the posttranslational regulation of *Saccharomyces cerevisiae* vacuolar hydrolases. Mol Cell Biol 1986; 6:2500-10.

85. Chang Y-H, Smith JA. Molecular cloning and sequencing of genomic DNA encoding aminopeptidase I from *Saccharomyces cerevisiae*. J Biol Chem 1989; 264:6979-83.

86. Bayliss ME, Prescott JM. Modified activity of *Aeromonas* aminopeptidase: metal ion substitutions and role of substrates. Biochemistry 1986; 25:8113-7.

87. Arfin SA, Kendall RL, Bradshaw RB. Eukaryotic methionyl aminopeptidases: Two classes of cobalt-dependent enzymes. Proc Natl Acad Sci USA 1995; 92:7714-8.

88. Miller CG, Strauch KL, Kukral AM et al. N-terminal methionine-specific peptidase in *Salmonella typhimurium*. Proc Natl Acad Sci USA 1987; 84:2718-22.

89. Bazan JF, Weaver LH, Roderick SL et al. Sequence and structure comparison suggest that methionine aminopeptidase, prolidase, aminopeptidase P, and creatinase share a common fold. Proc Natl Acad Sci USA 1994; 91:2473-77.

90. Moerschell RP, Hosokawa Y, Tsunawa S et al. The specificities of yeast methionine aminopeptidase and acetylation of amino-terminal methionine in vivo. J Biol Chem 1990; 265:19638-43.

91. Kendall RL, Bradshaw RA. Isolation and characterization of the methionine aminopeptidase from porcine liver responsible for the cotranslational processing of proteins. J Biol Chem 1992; 267:20667-73.

92. Asano Y, Nakazawa A, Kato Y et al. Properties of a novel D-ste-

reospecific aminopeptidase from *Ochrobactrum anthropi*. J Biol Chem 1989; 264:14233-39.

93. Enenkel C, Wolf DH. BLH1 codes for a yeast thiol aminopeptidase, the equivalent of mammalian bleomycin hydrolase. J Biol Chem 1993; 268:7036-43.

94. Nishimura T, Rhyu MR, Kato H. Purification and properties of aminopeptidase H from porcine skeletal muscle. Agric Biol Chem 1991; 55:1779-86.

95. Kunz J, Krause D, Kremer M et al. The 140 kDa protein of blood-brain barrier-associated pericytes is identical to aminopeptidase N. J Neurochem 1994; 62:2375-86.

96. Helene A, Beaumont A, Boques BP. Functional residues at the active site of aminopeptidase N. Eur J Biochem 1991; 196:385-93.

97. Orawski AT, Simmons WH. Purification and properties of membrane-bound aminopeptidase P from rat lung. Biochemistry 1995; 34:11227-36.

98. Orawski AT, Susz JP, Simmons WH. Aminopeptidase P from bovine lung: solublization, properties, and potential role in bradykinin degradation. Mol Cell Biochem 1987; 75:123-32.

99. Mock WL, Liu Y. Hydrolysis of picolinylprolines by prolidase. A general mechanism for the dual-metal ion containing aminopeptidases. J Biol Chem 1995; 270:18437-46.

100. Suda H, Aogi T, Takeuchi T. Inhibition of aminopeptidase B and leucine aminopeptidase by bestatin and its stereoisomer aminopeptidase: nonidentical, interacting metal-binding sites. Arch Biochem Biophys 1976; 177:196-200.

101. Kim H, Burley SK, Lipscomb WN. Re-refinement of the x-ray crystal structure of bovine lens leucine aminopeptidase complexed with bestatin. J Mol Biol 1993; 230:722-4.

102. Oettgen HC, Taylor A. Purification and characterization of leucine aminopeptidase from hog lens. Analyt Biochem 1995; 146:238-245.

103. Bernstein FC, Koetzle TF, Meyer EF Jr, Williams EF et al. The protein data bank: A computer-based archival file for macromolecular structures. J Mol Biol 1977; 112:535-542.

104. Abola, EE, Bernstein FC, Bryant SH et al. Protein data bank. In: Allen FH, Bergerhoff G, eds. Crystallographic Databases—Information Content, Software Systems. Bonn/Cambridge/Chester: Data Commission of Intl Union of Crystallography, 1987:107-132.

CHAPTER 3

METALLOBIOCHEMISTRY OF AMINOPEPTIDASES

Harold E. Van Wart

INTRODUCTION

The aminopeptidases (APs) constitute a diverse group of pro-teinases that share the property that they catalyze the hydroly-sis of amino acid residues from the N-terminus of peptide and protein substrates.[1,2] It has become clear that the great majority of APs are metalloenzymes. Although our understanding of the metal-lobiochemistry of the APs has generally lagged behind that of the metallo-endopeptidases and -carboxypeptidases, recent advances in our knowledge of the sequence, structure and function of a num-ber of prototypic APs have greatly clarified the picture. In this chapter, the APs will be discussed from the perspective of their metal dependencies with an emphasis on the role(s) of the active site metal atom(s) in catalysis. In particular, these roles will be related to the structures of the metal centers for prototypic APs.

CLASSIFICATION BY METAL CENTER

Three types of information have been accumulated that, taken collectively, allow a classification of APs according to their metal centers. The first type is functional information that demonstrates that the AP depends upon a metal atom for its activity. Most of-ten, this starts with inhibition studies with metal-chelating agents and is followed by more definitive studies which directly analyze the AP to identify the metal atom(s) present and to establish their

Aminopeptidases, edited by Allen Taylor. © 1996 R.G. Landes Company.

stoichiometry. Sometimes, replacement of the native metal atom with other metals is carried out to study its functional role in more detail or to introduce spectroscopic probes into the active site.

The second type of information is structural and follows from the elucidation of the x-ray structure of the AP. This gives the identity of the ligands to the metal(s) and also reveals the structure of the metal center. Once the structure of any prototypic metal center has been established, the identity and spacing of the metal ligands on the protein can be viewed as a "metal-binding motif". This then makes it possible to use a third type of information, protein sequences taken from databases, to assign known APs to certain classes and, potentially, to identify new APs. Based on the types of information described above, three prototypic classes of APs can presently be distinguished on the basis of their metal centers.

The first is a group of APs that probably contain a single zinc atom at the active site that is coordinated by two His and one Glu residues of the protein. The zinc-binding motif consists of the sequence $H-X_3-H-X_{19}-E$, where the Xs stand for the intervening residues in the "short" and "long" spacer sequences between the ligands. This metal-binding motif has actually been identified from the structure of thermolysin,[3] a zinc endopeptidase. Although there have been no x-ray structures solved for APs that contain this sequence, it is presumed by analogy with thermolysin to be the site of binding of a zinc atom in these APs. The sequences in the vicinity of the putative zinc ligands for nine APs are compared with that of thermolysin in Figure 3.1. In the nomenclature of Vallee and Auld,[4-8] this is termed a catalytic zinc site that is characterized by tetrahedral coordination to the protein with water occupying the fourth coordination site. In the metallopeptidase classification scheme of Rawlings and Barrett,[9] these APs fall into the alanyl AP, or M1, family of the HEXXH + E clan. This family includes the membrane alanyl APs, glutamyl AP, *E. coli* AP-N, and leukotriene A_4 hydrolase. Although all of these APs are presumed to contain one zinc atom, this has only been confirmed by direct analytical measurement for leukotriene A_4 hydrolase.

There are several APs that have been shown to contain one catalytically essential zinc atom per protein chain, but whose sequences are not yet known. These include rabbit kidney brush

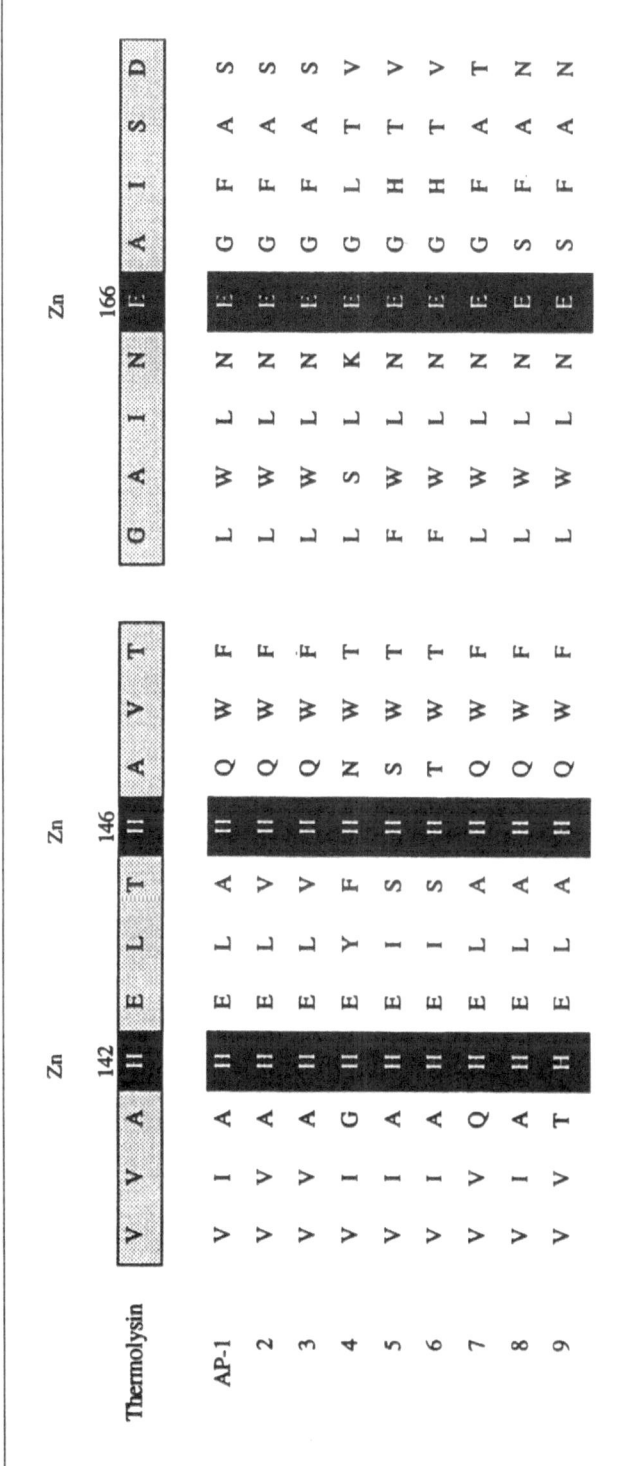

Fig. 3.1. Sequences surrounding the three zinc ligands (shaded boxes) in thermolysin and comparison with those of nine putative members of the one-zinc class of APs (or M1 family of metallopeptidases). These APs are presumed to bind zinc with the same residues as thermolysin. AP-1, human membrane alanyl AP; AP-2, mouse glutamyl AP; AP-3, human glutamyl AP; AP-4, E. coli AP-N; AP-5, human leukotriene A_4 hydrolase; AP-6, guinea pig leukotriene A_4 hydrolase; AP-7, yeast aminopeptidase yscII; AP-8, L. lactis lysyl AP; AP-9, L. delbrueckii lysyl AP. Data have been taken from reference 9.

border AP,[10] aminopeptidase Ey from hen's egg yolk,[11] and *S. griseus* AP.[12] One or more of these APs may also belong to the M1 family of alanyl APs. However, one or more may well possess metal-binding motifs that are distinct from that of the alanyl AP family described above. Thus, there may be more than one type of one-zinc AP. The crystal structure of *S. griseus* AP is currently under investigation and may reveal whether or not this is the case.[13] This AP is very interesting because it is activated by calcium ions. At the present time, not enough data are available to classify these one-zinc APs.

The second class of APs is distinguished by the presence of two intrinsic zinc atoms bound closely together with a bridging carboxylate ligand. Vallee and Auld refer to this as a cocatalytic zinc site.[6,7] The x-ray structures of two distinct types of APs have been elucidated that contain this type of two-zinc center. The first is the hexameric bovine lens AP (blLAP)[14-18] whose structure establishes that the two zinc ions are bound by Asp, Glu and Lys ligands with the $K-X_4-D-X_{17}-D-X_{58}-D-X-E$ motif. This motif is also found in the sequences of four other APs, including *E. coli* AP-A (Fig. 3.2). Although its sequence has not been reported, particulate AP from porcine kidney has been shown to contain two catalytically essential zinc atoms[19] and may belong to this class as well. This family has been termed the leucine aminopeptidase or M17 family of metallopeptidases.[9]

A second type of two-zinc AP is the monomeric AP from *Aeromonas proteolytica* (ap-AP). The x-ray structure of this AP has been solved, and it has been shown to have a similar type of two-zinc center.[20] The structure, however, shows that the zinc atoms are bound by His, Asp, and Glu ligands with the distinct metal-binding motif $H-X_{19}-D-X_{34}-E-X_{26}-D-X_{76}-H$[21] (Fig. 3.2). No other APs with the ap-AP zinc binding motif are currently known. Thus, while the two-zinc centers of blLAP and ap-AP are similar in structure, they are formed by two different sets of ligands with markedly different spacings.

There are some APs that contain only one tightly bound zinc ion on isolation, but that contain a second metal-binding site that can be occupied by subsequent incubation with certain metal ions. Soluble porcine kidney AP (PK-AP) is almost certainly a member of the M17 class of two-zinc leucine aminopeptidases described above, but the zinc atom at one site binds more weakly than that

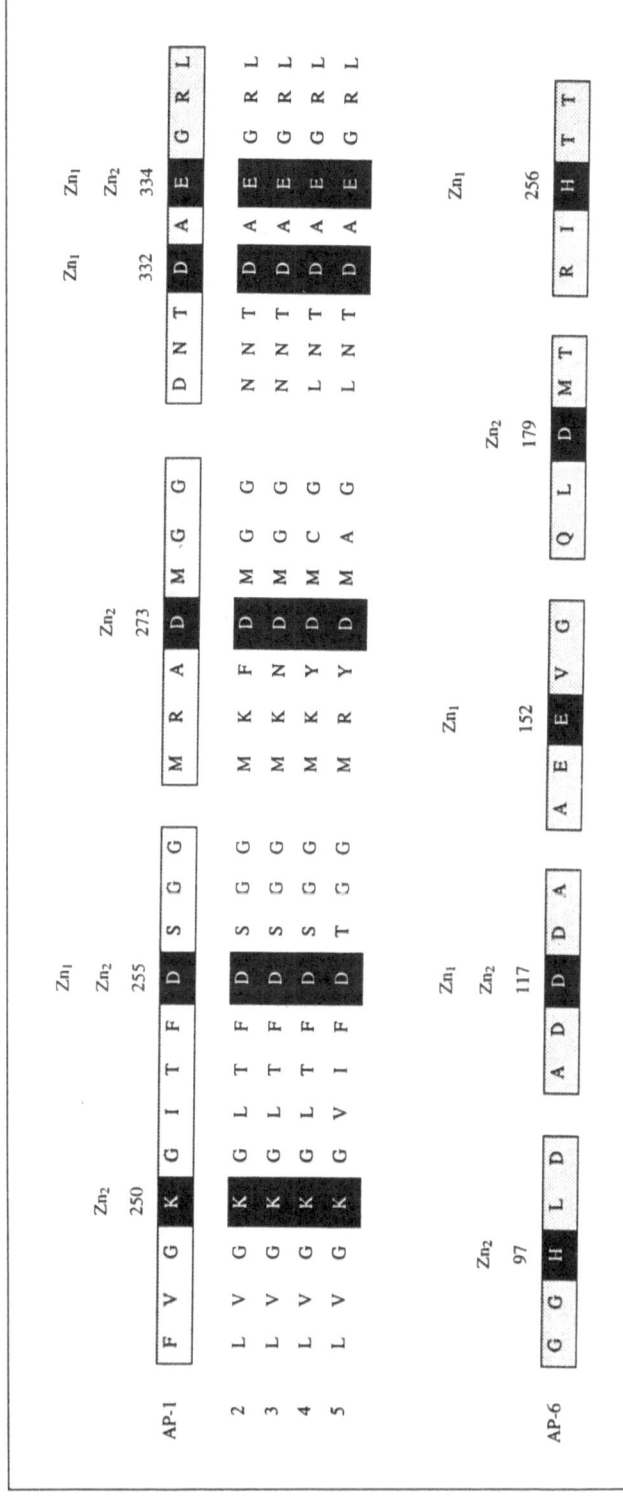

Fig. 3.2. *Sequences surrounding the zinc ligands (shaded boxes) in the two-zinc class of APs (leucyl APs or M17 family of metallopeptidases). The first type of AP in this class is exemplified by blLAP, or AP-1, shown at the top of this figure, whose crystal structure establishes that the shaded residues are ligands to the two zinc atoms. The sequences of four other APs that are presumed to have the same zinc ligands are also shown. AP-2, A. thaliana leucyl AP; AP-3, potato leucyl AP; AP-4, E. coli AP; AP-5, R. prowazekii leucyl AP. Data have been taken from reference 9. The second type of AP in this class is AP-6 or ap-AP, shown at the bottom of this figure, whose crystal structure reveals the shaded zinc ligands: The sequence data are from reference 53. Although both types of two-zinc APs have similar metal centers, they have different ligands and little sequence homology surrounding them. The protein ligands which bind the two nonequivalent zinc atoms, denoted Zn_1 and Zn_2, in both types of APs are indicated.*

at the other,[22] and the isolated enzyme contains only one zinc per subunit. On this basis, it would be mistakenly classified as a one-zinc AP. Thus, the relative binding constants for the two sites in the two-zinc APs may vary and cause them to be confused with the one-zinc class. The situation is similar for human liver AP[23] and *B. subtilis* AP,[24] both of which have been shown to contain one zinc atom per protein chain on isolation. However, incubation of these APs with cobalt leads to activation of the enzymes due to binding of one atom of cobalt in addition to the intrinsic zinc. These enzymes may constitute their own separate type of two-metal atom AP. However, this classification will have to await the determination of their sequences and crystal structures.

The third class of APs contains a two-cobalt cluster that shares many similarities with the two-zinc APs. The prototypic member of this class is *E. coli* methionyl AP whose x-ray structure[25] shows that the cobalt atoms are ligated by Asp, Glu, and His residues of the protein with the sequence $D-X_{10}-D-X_{62}-H-X_{32}-E-X_{30}-E$. Six other APs have sequences that contain this motif (Fig. 3.3). This family has been termed the methionyl aminopeptidase or M24 family.[9] In the remainder of this chapter, discussion will be limited to the three classes of APs for which there is structural information.

THE M1 FAMILY OF ONE-ZINC APs

There is relatively little known about this class of APs. The only member for which accurate zinc analyses have been performed is leukotriene A_4 hydrolase (LTA$_4$ hydrolase) which has been shown to contain one zinc atom per molecule[26,27] (see chapter 6). This enzyme has two activities. The one for which it is named is the hydrolysis of the epoxide LTA$_4$ to produce LTB$_4$. However, it has also been shown to have aminopeptidase activity.[28-30] Both activities are abolished on removal of the zinc, but can be restored by addition of stoichiometric amounts of zinc or cobalt ions.[26,29,30] The crystal structure of LTA$_4$ hydrolase has not been solved, but the identity of the metal ligands in the mouse enzyme has been studied by site-directed mutagenesis.[31] Mutation of each of the three putative zinc ligands (H295Y, H299Y, and E318Q; numbering of residues as found in LTA$_4$ hydrolase, not thermolysin) simultaneously abolishes both activities of the enzyme and lowers its zinc content to insignificant levels. This is strong circumstantial support for the zinc binding site shown in Figure 3.1.

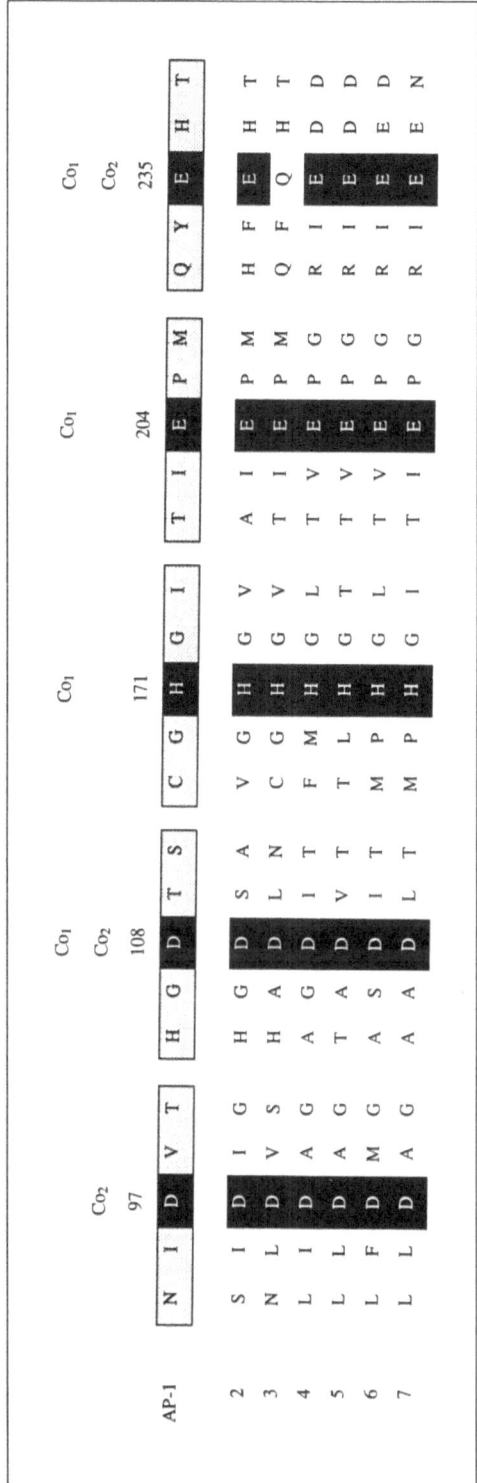

Fig. 3.3. Sequences surrounding the cobalt ligands (shaded boxes) in the two-cobalt class of APs (methionyl AP or M24 family). The crystal structure of AP-1, E. coli methionyl AP, establishes that the shaded residues are the cobalt ligands. The six other APs whose sequences are shown are presumed to bind the two cobalt atoms in a similar fashion. AP-2, B. subtilis methionyl AP; AP-3, yeast methionyl AP; AP-4, E. coli X-Pro AP; AP-5, S. lividans X-Pro aminopeptidase; AP-6, human X-Pro dipeptidase; AP-7, E. coli X-Pro dipeptidase. The protein ligands which bind the two nonequivalent cobalt atoms, denoted Co$_1$ and Co$_2$, are indicated.

Although there are no crystal structures available for any of the M1 family of APs, the zinc site is expected to resemble that of the prototypic catalytic zinc site found in thermolysin. In this site, the zinc atom is coordinated to the two His and Glu ligands in an approximately tetrahedral fashion with the fourth site occupied by water,[3] as shown in the schematic in Figure 3.4. The metal binding motif in this class of peptidases actually has the partial sequence HEXXH, where the conserved Glu is an essential catalytic residue. The function of a catalytic zinc site is to activate water for use in the peptide bond breaking step, and both this Glu residue and the zinc atom participate in this activation.

A variety of catalytic mechanisms is possible, and a leading proposal is one in which the carbonyl oxygen atom of the scissile peptide bond coordinates to the zinc atom in a five-coordinate complex.[3] The zinc then functions as a Lewis acid to activate this group for nucleophilic attack by the zinc-bound water molecule. The catalytic Glu residue acts as a general base to facilitate this attack by hydrogen bonding to the zinc-bound water. This type of mechanism has been proposed for both zinc endopeptidases as well as zinc carboxypeptidases and is also a plausible mechanism for the one-zinc APs. It should be noted that a catalytic role has been confirmed for the Glu-296 residue of mouse LTA4 hydrolase.[32] Mutagenesis studies have demonstrated that the E296Q and E296A

Fig. 3.4. Schematic showing the structure of the zinc site for the M1 family of APs based on the presumption that this site is similar to that in thermolysin.

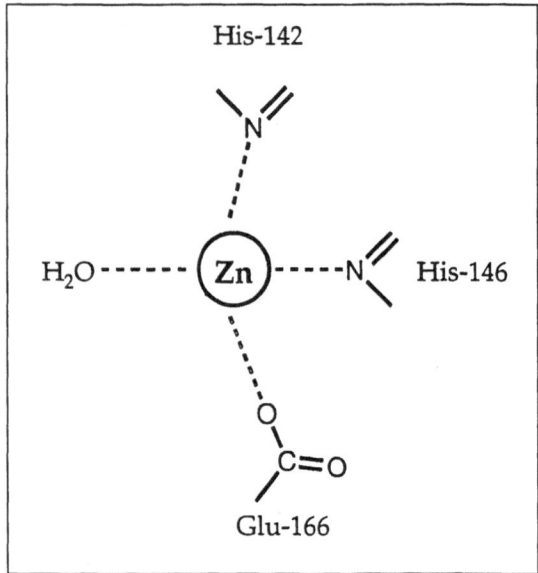

mutations, which do not alter the zinc content of the enzyme, abolish aminopeptidase activity but not epoxide hydrolase activity. These results confirm that this Glu residue is playing a catalytic role in the aminopeptidase reaction.

THE TWO-ZINC CLASS OF APs

By far the most is known about the two-zinc class of APs. With respect to the first type of two-zinc site distinguished at the top of Figure 3.2, a considerable amount of work has been carried out on blLAP and PK-AP. Thus, discussion will be limited to these two enzymes. BlLAP is a hexameric AP consisting of identical 54 kDa subunits that was originally shown by Carpenter and Vahl[33] to contain two atoms of catalytically essential zinc per subunit. Removal of the zinc by dialysis gives an inactive enzyme that can be fully reconstituted by binding of two zinc ions per subunit. The apoenzyme cannot be reactivated by copper, calcium, magnesium, or manganese ions alone. These workers also established that the two zinc sites were functionally and chemically nonequivalent and this has led to a range of terms that have been used to describe the two zinc sites. Because of the significance of these classifications to the subject of this chapter and as an aid to the reader, these terms are summarized in Table 3.1.

The two zinc sites will be designated as site 1 and site 2. Incubation of the two-zinc enzyme with manganese, magnesium, or cobalt ions leads to rapid exchange of the zinc atom in site 1 by these metals.[34] Accordingly, this site has been designated the rapidly exchanging site. Since this exchange is accompanied by activation of the enzyme, this site has also been called the activation

Table 3.1. Terms and designations used to differentiate the two zinc sites in blLAP

Site 1	Site 2	Ref.
Rapidly exchanging	Slowly exchanging	34
Activation	Specificity	33
Activation	Structural	35
Regulatory	Catalytic	22
Zn*	Zn	6
Zn488	Zn489	17

site.[33,35] The zinc atoms in both sites can be exchanged with cobalt by more prolonged dialysis with retention of activity.[34,35] Since the exchange of cobalt into site 2 takes place more slowly, it has been termed the slowly exchanging site.[34] Occupancy of this site by only zinc and cobalt gives an active enzyme. Thus, this site has also been termed the specificity site[33] or structural site.[35] Although the identity of the metal atom in site 1 most markedly affects activity, the metal in site 2 is important as well.[34] It should be appreciated that the terms described above originated before it was appreciated that the two zinc atoms were located in close proximity to each other and probably act interdependently (see below).

Another closely related two-zinc AP is PK-AP. Although sequence data are not available for this enzyme, it is known to be immunologically and functionally very similar to blLAP[36,37] and almost certainly has a similar ligand set. PK-AP is also a hexamer consisting of identical 54 kDa subunits,[38] but differs from blLAP in that each subunit contains only one tightly bound zinc atom.[22,39] Unlike blLAP, the one-zinc form of PK-AP is active.[22] For this reason, the zinc atom in the one-zinc form was termed the catalytic zinc. Smith and Spackman showed that PK-AP is activated by manganese and magnesium ions, but is inhibited by zinc, nickel, copper, cadmium, and mercury ions.[40] While it was originally hypothesized that the activation by manganese was due to replacement of the catalytic zinc by manganese,[39] it has been shown instead that the effect of all of these metals atoms can be attributed to their binding to a second distinct site in addition to the site occupied by the catalytic zinc.[22] Because these metal ions modulate the activity of the one-zinc form of PK-AP (either activate or inhibit), the site to which they are bound has been termed the regulatory site.[22] This corresponds to site 1 for blLAP, while the catalytic site referred to above corresponds to site 2. Thus, the metal binding behaviors of blLAP and PK-AP are very similar with the exception that the binding of metal ions to site 1 in PK-AP is weaker, and the one-zinc form of PK-AP is active.

A variety of kinetic measurements has been used to describe the activities of the metallo-APs discussed above. In particular, the effects of metal substitutions at the two different sites on the steady-state kinetic parameters k_{cat} and K_m have been studied for both bl- and PK-AP. The activation of blLAP by magnesium and manganese was originally reported to be due predominantly to an increase in k_{cat} for hydrolysis of the substrate Leu-p-nitroanilide.[33] A subsequent study with this substrate suggested that the replacement of zinc by cobalt at site 2 influenced activity predominantly by lowering K_m.[35] A more thorough study of the independent effects of substitution of cobalt and magnesium for zinc at site 1, and of cobalt for zinc at site 2 with three different substrates (Leu-p-nitroanilide, Leu-NH$_2$, and Leu-p-anisidine) has since been carried out.[34] This study shows that substitutions at both sites affect both kinetic parameters and that the trends depend upon the identity of the substrate.

A similar situation with respect to metal ion substitution at site 1 exists with PK-AP. Variation of the metal atom at this site was originally shown to modulate the activity of the enzyme toward Leu-p-nitroanilide by altering k_{cat} and leaving K_m unchanged.[22] However, a subsequent study using the substrate Leu-Gly-NHNH-Dns showed that metal substitutions at site 1 affect both k_{cat} and K_m.[41,42] The observation that use of a substrate having true amino acids on both sides of the scissile bond gives different results than those obtained with the more artificial chromogenic p-nitroanilide substrates suggests that caution be exercised in over-interpreting the function of the individual metal sites using nonphysiological substrates.[42a]

PK-AP has been shown to be particularly resistant to denaturation by organic cosolvents.[43] This has made it possible to carry out low-temperature stopped flow studies of the presteady-state reaction with Leu-Gly-NHNH-Dns.[44] These studies show that the reaction proceeds through the formation of two distinct enzyme-substrate complexes for the enzymes with nickel, copper and zinc in site 1, while only a single intermediate could be detected for the species with magnesium and manganese in this site. This shows that the metal atom in site 2 also influences the presteady-state kinetics of this reaction.

Solution of the crystal structure of blLAP and its complexes with bestatin,[14,15] amastatin,[16] and Leu phosphonic acid[18] provided a detailed view of the structure of its two-zinc site. The two zinc atoms are approximately 2.9 Å apart and are bound by different ligands. A schematic of the coordination spheres of the two zinc atoms is shown in Figure 3.5A. The zinc atom at the bottom of this figure is coordinated by the $O^{\delta 1}$ atom of Asp-255, the carbonyl oxygen and $O^{\delta 2}$ atoms of Asp-332, and the $O^{\epsilon 2}$ atom of

Glu-334. The second zinc atom, shown at the top, is bridged to the first by the $O^{\epsilon 1}$ atom of Glu-334 and the $O^{\delta 1}$ atom of Asp-255. It is also ligated by the N^ζ atom of Lys-250 and the $O^{\delta 1}$ atom of Asp-273.[16] In the original structural determination of the two-zinc form of blLAP,[14,15] it was not immediately obvious which of these two zinc atoms were bound at sites 1 and 2. Because the temperature factor for the zinc atom shown at the bottom of Figure 3.5 was considerably smaller than that for the zinc atom shown at the top, the bottom zinc atom was originally assigned as the slowly exchanging site 2 zinc atom.[15] The assignment, however, has now been reversed. The structure of the metallohybrid with zinc in site 2 and magnesium in site 1 has been elucidated recently by Kim and Lipscomb.[17] These workers observed a lower electron density for the metal atom in the bottom site shown in Figure 3.5A. Since magnesium is a weaker x-ray scatterer than zinc, this identifies the bottom site as site 1. These authors referred to the site 1 zinc atom as Zn488 and the site 2 zinc atom as Zn489. This two-zinc site has been termed a cocatalytic site by Vallee and Auld.[6,7] In their terminology, the catalytic zinc is designated as Zn and the other zinc as Zn*.

With the studies reviewed above in hand, it is possible to integrate the functional and structural data to formulate a mechanism of action for blLAP (see chapter 2). Based on the structures of the complexes with bestatin, amastatin, and Leu phosphonic acid, interactions believed to be important for substrate hydrolysis have been identified and a number of potential mechanisms proposed.[18] While the details vary, these mechanisms share several common features. First, the carbonyl group of the scissile peptide bond is activated by coordination to one of the zinc atoms, probably that at site 1. Second, the attacking nucleophile is a water molecule (or hydroxide ion) that is bound to one or both of the zinc atoms. This feature of the mechanism has been difficult to pin down because no unambiguously defined zinc-bound water molecules have been observed in any of the crystal structures. Third, the terminal αNH_2 group of the substrate binds to the zinc atom at site 2. Fourth, the attack of the zinc-bound water on the substrate carbonyl group leads to the formation of a *gem*-diolate intermediate. Fifth, Arg-336 and Lys-262 may be facilitating catalysis by stabilization of the zinc-bound hydroxide, stabilization of the *gem*-diolate intermediate, or protonation of the amino group of the departing

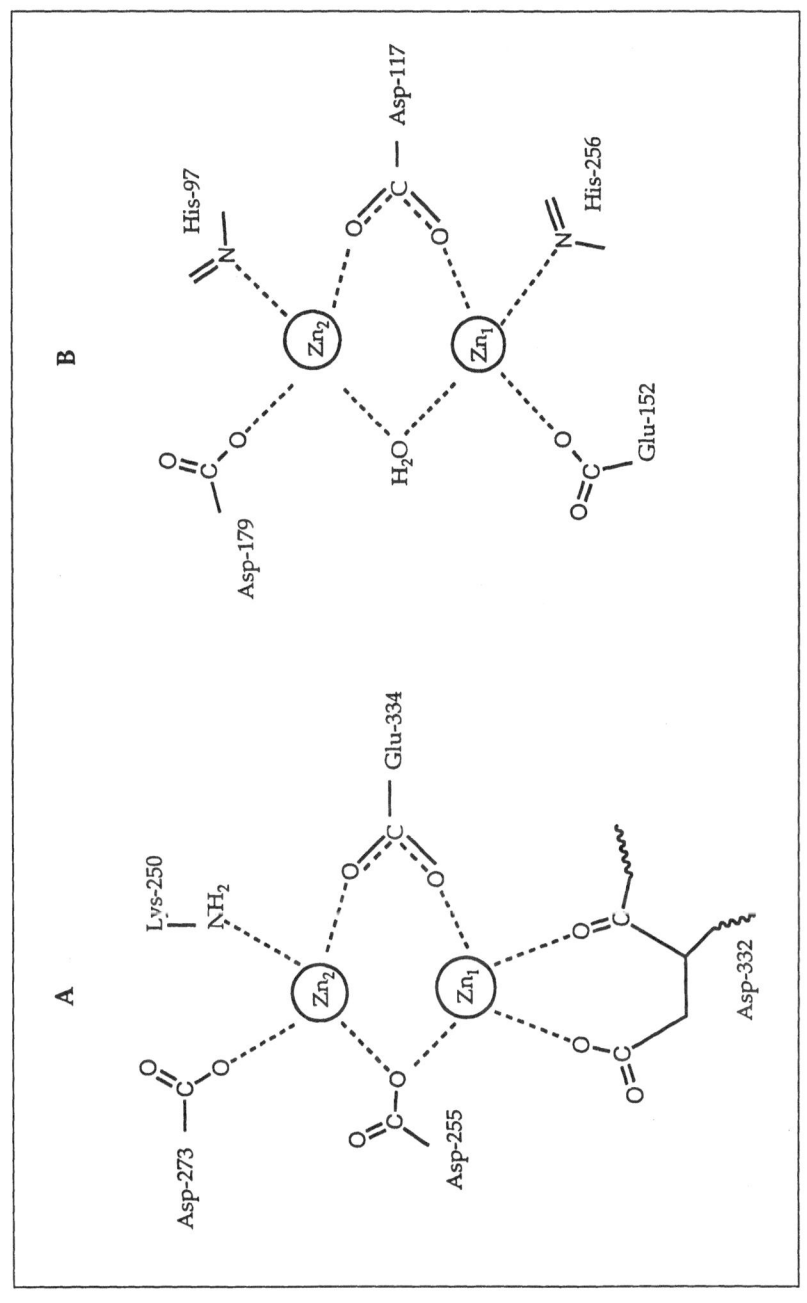

Fig. 3.5. Schematic showing the structures of the two-zinc sites in (A) blLAP[14-18] and (B) ap-AP.[20]

P_1' residue. The close association of the two zinc atoms suggests that they are acting in concert to catalyze the reaction and that they both function in both the binding and catalytic steps. Thus, the original concept that each could be assigned the independent roles delineated in Table 3.1 is oversimplified.

Ap-AP is a monomeric AP with a subunit molecular weight of 29 kDa that contains two atoms of zinc per protein chain.[45] Unlike blLAP and PK-AP, ap-AP is not activated by magnesium or manganese ions. The apoenzyme is inactive, but can be fully reactivated by addition of zinc. Reconstitution with cobalt, nickel, or copper results in species that are 5.0-, 9.8- and 10-fold more active (as reflected by the k_{cat}/K_m value for hydrolysis of Leu-p-nitroanilide) respectively, than the native enzyme.[47] Titrations with all four of these metal ions demonstrate that full activity is restored by binding of one equivalent of metal ion. Magnesium, manganese, and calcium ions all fail to reactivate the apoenzyme.

In contrast to the straightforward results obtained when apo-ap-AP is reconstituted with two equivalents of the same metal ion, surprising results are obtained when single equivalents of zinc, cobalt, nickel, and copper are added in different orders. Addition of a single equivalent of either nickel or copper followed by one equivalent of zinc gives 52- and 73-fold activation, respectively, relative to the apoenzyme reconstituted with two equivalents of zinc.[47] However, if the order of additions is reversed, the activations are only 3.9- and 3.4-fold, respectively. These results demonstrate that occupancy of the first site is sufficient for catalytic activity. Accordingly, site 1 for ap-AP was termed the catalytic site. The results also show that the identity of the metal atom in the second site markedly influences activity. This site has been called the regulatory or structural site. The pattern of activation observed in these experiments shows that the level of activity of each metallohybrid is a characteristic of each pair of metal ions in the two sites. Thus, the effect of having one metal ion in site 2 is different depending on the identity of the metal ion in site 1. Thus, the two metal sites are nonidentical and interacting.

The effect of the occupancy of these two sites by different metal ions on the individual kinetic parameters for substrate hydrolysis has also been investigated. Bayliss and Prescott studied the hydrolysis of Leu-, Ala-, and Val-p-nitroanilide by all 16 possible combinations of zinc, cobalt, nickel, and copper at the two metal binding sites.[48] Very large variations in the kinetic parameters were observed for the different hybrids with the different substrates with the values of k_{cat} varying over 1800-fold, K_m by more than 300-fold and k_{cat}/K_m by more than three orders of magnitude. The identity of the substrate was quite important with Leu-p-nitroanilide the best for all metallohybrids. As with blLAP- and PK-AP, these studies showed that neither of the two individual metal sites exclusively controls the values of either k_{cat} or K_m. Both parameters are governed in a complex manner by the identity of the metal ions in both sites. The complex relationship between the identity of the metal ion at each site, the structure of the substrate, and the two different kinetic parameters reflect the close cooperation of the two metal atoms in catalyzing the hydrolysis of the substrate.

Although ap-AP differs from blLAP in many ways, an attempt has been made to compare the two metal binding sites based on the functional data described above.[48] Since activity is restored to apo-ap-AP when one equivalent of metal binds to site 1, it has been suggested that this is the analog of the catalytic, slowly exchanging site 2 in blLAP. Likewise, the stimulatory effect of having the second site in ap-AP occupied by a metal ion suggested that it was the analog of the rapidly exchanging site 1 in blLAP. The apparent opposite designation of sites with similar functions in ap-AP and blLAP has arisen because of the different methods of preparation of the metallo-ap-AP and -blLAP species. For ap-AP, this was accomplished by titration of the apoenzyme with metal ions such that site 1, the slowly exchanging (and presumably the stronger binding) site, was occupied first, and site 2, the rapidly exchanging (and presumably the weaker binding) site, was occupied second. In contrast, the preparation of the blLAP hybrids was accomplished by displacement of the native zinc atoms in the two sites such that the first site that was exchanged, site 1, corresponded to the weaker binding site, and the second site that was exchanged, site 2, corresponded to the stronger binding site.

The x-ray crystal structure of the two-zinc form of ap-AP has recently been reported.[20] Although there is an insignificant level of homology between ap-AP and blLAP, the carboxyl terminal domains of the two APs have similar folds. The metal center of ap-AP is located in a loop region near the carboxyl-terminal edge of two parallel strands. The distance between the two zinc atoms is 3.5 Å, somewhat greater than in blLAP. Based on the functional nonequivalence of the two metal ion sites discussed above, it was surprising to find that the two zinc atoms appear structurally equivalent. A schematic of the structure of the zinc center of ap-AP is shown in Figure 3.5B in which an attempt has been made to orient the zinc atoms and ligands to highlight the similarities with the blLAP two-zinc center shown in Figure 3.5A. Both zinc ions in ap-AP have approximately tetrahedral coordination with the carboxyl group of Asp-117 and the oxygen atom of water serving as bridging ligands. It is not clear which of the zinc atoms in the crystallographically determined structure correspond to the site 1 and 2 zinc atoms discussed above. The zinc atoms shown at the bottom and top are labeled Zn_1 and Zn_2, respectively, for consistency with the crystallographic nomenclature only. The ligands to Zn_1 are the $O^{\epsilon 1}$ atom of Glu-152, the $N^{\epsilon 2}$ atom of His-256, the $O^{\delta 2}$ atom of Asp-117, and the oxygen atom of the bridging water (or hydroxide). The ligands to Zn_2 are the $O^{\delta 1}$ atom of Asp-117, the O atom of the bridging water, the $N^{\epsilon 2}$ atom of His-97, and the $O^{\delta 1}$ atom of Asp-179. There also appear to be weaker interactions between the $O^{\epsilon 2}$ atom of Glu-152 and Zn_1, and the $O^{\delta 2}$ atom of Asp-179 and Zn_2.

Although both ap-AP and blLAP contain binuclear zinc centers that share an overall similarity, the metal centers of the two APs differ in a number of ways. First, ap-AP contains a water molecule within the first coordination sphere of the zinc ions, while in the structure of blLAP this is not clearly indicated. The direct observation of zinc-bound water in ap-AP provides support for a role of one or both of the zinc atoms in activating water as part of the catalytic mechanism. Second, the bridging ligands and zinc-zinc distances differ in the two APs. Third, the ligand sets are more symmetrical in ap-AP, which has one His coordinated to each zinc, while blLAP has a Lys coordinated only to Zn_2. This makes it difficult to understand the basis for the functional nonequivalence of the two sites in ap-AP as well as to speculate as to which of the sites 1 and 2 identified in the functional studies corresponds to the Zn_1 and Zn_2 atoms identified in the x-ray structure. The differences in the behavior of the two zinc atoms may be the consequence of subtle differences in their coordination spheres. Last, the active site of ap-AP does not contain residues analogous to Lys-262 and Arg-336 in blLAP that have been hypothesized to play catalytic roles. Thus, it is likely that there are significant mechanistic differences between ap-AP and blLAP.

THE M24 FAMILY OF METHIONYL APs

The methionyl AP family is a very important class of enzymes that removes the N-terminal Met residue from newly synthesized proteins. This is a prerequisite for a number of possible co- or post-translational modifications which produce modified proteins with a variety of N-termini. Although these APs appear to be metalloenzymes, our knowledge of their metal content and function is still very sketchy. The enzymes from *E. coli*,[49] *S. cerevisiae*,[50] *S. typhimurium*[51] and porcine liver[52] have been isolated and studied (see chapters 1 and 2). All of these enzymes are inhibited by metal chelating agents, but none have been analyzed for their native metal content or stoichiometry. Thus, the identity of the native metal has not been unequivocally established. However, all of these APs are stimulated by cobalt and are either unaffected or inhibited by zinc, magnesium and manganese. Yeast apo-AP has been prepared by dialysis against EDTA. The apoenzyme could be reactivated by cobalt, but not magnesium, manganese, copper, iron or zinc.[50] The stimulatory effect of cobalt on this family of APs has lead to the presumption that they are cobalt enzymes.

The x-ray structure of *E. coli* AP (ec-AP) has recently been elucidated.[25] ec-AP is a monomeric AP with a molecular weight of 29 kDa. Crystals of the enzyme were grown in the presence of cobalt, and the structure of the enzyme reveals the presence of a binuclear metal center. The overall fold of the enzyme is novel and distinct from blLAP and ap-AP. The metal center is located on one side of a central β-sheet. A schematic showing the protein ligands is shown in Figure 3.6. The distance between the two cobalt ions, designated in the structure as Co_1 and Co_2, is 2.9 A. The two ions are bridged by the carboxylate groups of Asp-108 and Glu-235. The full coordination sphere of Co_1 consists of the $O^{\delta 2}$ atom of Asp-108, the $N^{\epsilon 2}$ atom of His-171 and the $O^{\epsilon 2}$ atoms of Glu-204 and Glu-235. The Co_2 atom is ligated by the $O^{\delta 1}$ and $O^{\delta 2}$ atoms of Asp-97, the $O^{\delta 1}$ atom of Asp-108, and the $O^{\epsilon 1}$ atom of Glu-235. All of these ligands are conserved in the sequences shown in Figure 3.3, except for *B. subtilis* AP in which Glu-235 is reported to be Gln. The arrangement of the four ligands

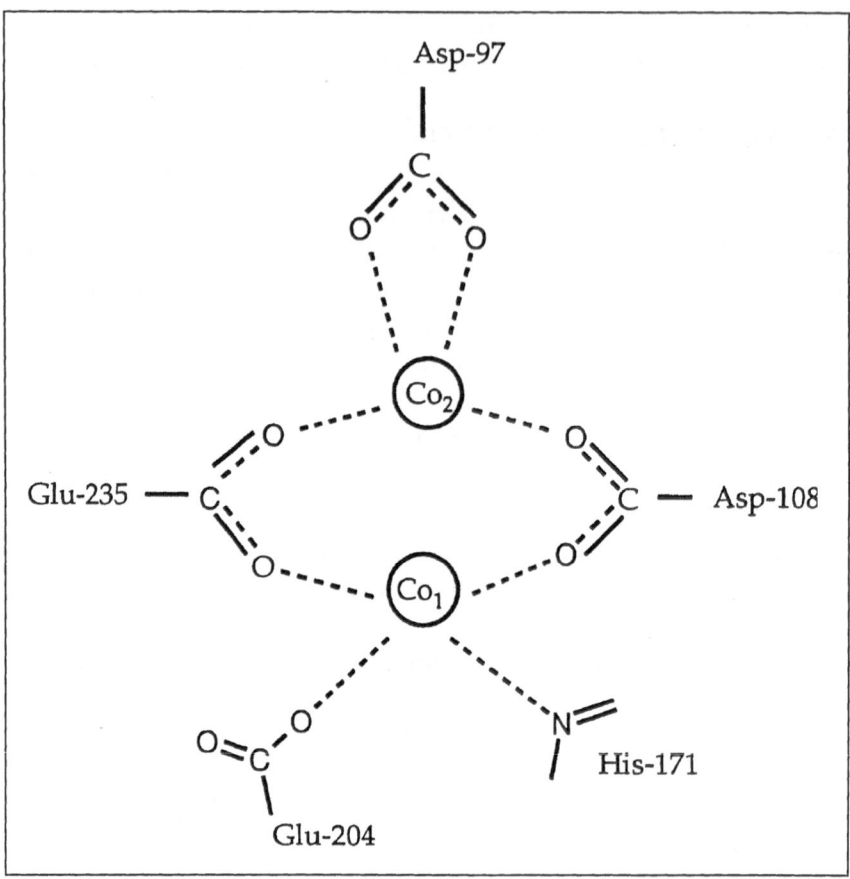

Fig. 3.6. Schematic showing the structure of the two-cobalt site in ec-AP.[25]

and second cobalt atom around each cobalt is approximately octahedral, and it has been suggested that water molecules could be
bound to the sixth sites of each cobalt. ec-AP also does not contain Lys or Arg residues in the active site that could play the catalytic roles suggested in blLAP. It does, however, have His-178 as
an active site residue that could serve an analogous catalytic role.

While ec-AP resembles the bl- and ap-APs in terms of its substrate specificity and reliance on a two-metal center, it differs markedly from these other two-metal APs in its overall protein fold,
the structure of the binuclear center, and, presumably, the identity of the catalytic metal. The functional consequences of changing the active site metal atom certainly differs markedly from that
in bl- and ap-AP. Thus, ec-AP constitutes an evolutionarily distinct enzyme that probably has a different mechanism. Too little

data are currently available for this class of AP to speculate on a mechanism, although a role for the binuclear metal center in activating the substrate and water is likely.

CONCLUSIONS

While our understanding of the role of the metal in AP function has lagged behind that of other classes of metallopeptidases, tremendous strides have now been made for the prototypic APs described in this and other chapters in this book. It appears that the diversity that the APs exhibit in their functions carries over to the structures of the individual proteins and their metal centers. Thus, it is likely that the number of classes of APs defined in this chapter will continue to increase as more primary and tertiary structural information becomes available. If so, the metallobiochemistry of AP will continue to be an important and active field for study.

ACKNOWLEDGMENTS

The author would like to thank Dr. Vera Imper for her valuable assistance in the preparation of this manuscript. This work was supported by National Institutes of Health grant GM27276 from the U.S. Public Health Service.

REFERENCES

1. Taylor A. Aminopeptidases: structure and function. FASEB J 1993; 7:290-98.
2. Taylor A. Aminopeptidases: toward a mechanism of action. Trends Biochem Sci 1993; 18:167-72.
3. Matthews BW. Structural basis of the action of thermolysin and related zinc peptidases. Acc Chem Res 1988; 21:333-40.
4. Vallee BL, Auld DS. Active-site zinc ligands and activated H_2O of zinc enzymes. Proc Natl Acad Sci USA 1990; 87:220-24.
5. Vallee BL, Auld, DS. Zinc coordination function and structure of zinc enzymes and other proteins. Biochem 1990; 24:5647-59.
6. Vallee BL, Auld DS. New perspective on zinc biochemistry: cocatalytic sites in multi-zinc enzymes. Biochem 1993; 32: 6493-6500.
7. Vallee BL, Auld DS. Cocatalytic zinc motifs in enzyme catalysis. Proc Natl Acad Sci USA 1993; 90:2715-18.
8. Vallee BL, Auld DS. Acc Chem Res 1993; 26:543-51.
9. Rawlings ND, Barrett AJ. Evolutionary families of metallopeptidases. Methods Enzymol 1995; 248:183-228.
10. Kerr MA, Kenny AJ. The molecular weight and properties of a neutral metallo-endopeptidase from rabbit kidney brush border.

Biochem J 1974; 137:489-95.

11. Tanaka T, Ichishima E. Molecular properties of aminopeptidase Ey as a zinc metalloenzyme. Int J Biochem 1993; 25:1681-88.

12. Spungin A, Blumberg S. *Streptomyces griseus* aminopeptidase is a calcium-activated zinc metalloprotein. Eur J Biochem 1989; 183:471-77.

13. Almog O, Greenblat HM, Spungin A et al. Crystallization and preliminary crystallographic analysis of *streptomyces griseus* aminopeptidase. J Mol Biol 1993; 230:342-44.

14. Burley SK, David PR, Taylor A et al. Molecular structure of leucine aminopeptidase at 2.7 A resolution. Proc Natl Acad Sci USA 1990; 87:6878-82.

15. Burley SK, David PR, Sweet RM et al. Structure determination and refinement of bovine lens leucine aminopeptidase and its complex with bestatin. J Mol Biol 1992; 224:113-140.

16. Kim H, Lipscomb WN. X-ray crystallographic determination of the structure of bovine lens leucine aminopeptidase complexed with amastatin: formulation of a catalytic mechanism featuring a *gem*-diolate transition state. Biochem 1993; 32:8465-78.

17. Kim H, Lipscomb WN. Differentiation and identification of the two catalytic metal binding sites in bovine lens leucine aminopeptidase by x-ray crystallography. Proc Natl Acad Sci USA 1993; 90:5006-10.

18. Strater N, Lipscomb WN. Transition state analogue L-leucine-phosphonic acid bound to bovine lens leucine aminopeptidase: x-ray structure at 1.65 A resolution in a new crystal form. Biochem 1995; 34:9200-10.

19. Lehky P, Stein EA. Enzymatic determination of zinc below one part per billion. Anal Chim Acta 1974; 70:85-93.

20. Chevrier B, Schalk C, D'Orchymont H et al. Crystal structure of *aeromonas proteolytica* aminopeptidase: a prototypical member of the cocatalytic zinc enzyme family. Structure 1994; 2:283-91.

21. Guenet C, Lepage P, Harris BA. Isolation of the leucine aminopeptidase gene from *aeromonas proteolytica*. J Biol Chem 1992; 267:8390-95.

22. Van Wart HE, Lin SH. Metal binding stoichiometry and mechanism of metal ion modulation of the activity of porcine kidney leucine aminopeptidase. Biochem 1981; 20:5682-89.

23. Garner CW, Behal FJ. Human liver aminopeptidase. role of metal ions in mechanism of action. Biochem 1974; 13:3227-33.

24. Wagner FW, Ray LE, Ajabnoor MA et al. *Bacillus subtilis* aminopeptidase: purification, characterization and some enzymatic properties. Arch Biochem Biophys 1979; 197:63-72.

25. Roderick SL, Matthews BW. Structure of the cobalt-dependent methionine aminopeptidase from *escherichia coli*: a new type of proteolytic enzyme. Biochem 1993; 32:3907-12.

26. Haeggstrom JZ, Wetterholm A, Shapiro R et al. Leukotriene A_4

hydrolase: a zinc metalloenzyme. Biochem Biophys Res Commun 1990; 172:965-70.

27. Toh H, Minami M, Shimizu T. Molecular evolution and zinc ion binding motif of leukotriene A$_4$ hydrolase. Biochem Biophys Res Commun 1990; 171:216-21.

28. Minami M, Ohishi N, Mutoh H et al. Leukotriene A$_4$ hydrolase is a zinc-containing aminopeptidase. Biochem Biiophys Res Commun 1990; 173:620-26.

29. Wetterholm A, Medina JF, Radmark O et al. Recombinant mouse leukotriene A$_4$ hydrolase: a zinc metalloenzyme with dual enzymatic activities. Biochim Biophys Acta 1991; 1080:96-102.

30. Haeggstrom JZ, Wetterholm A, Vallee BL et al. Leukotriene A$_4$ hydrolase: an epoxide hydrolase with peptidase activity. Biochem Biophys Res Commun 1990; 173:620-26.

31. Medina JF, Wetterholm A, Radmark O et al. Leukotriene A$_4$ hydrolase: determination of the three zinc-binding ligands by site-directed mutagenesis and zinc analysis. Proc Natl Sci USA 1991; 88:7620-24.

32. Wetterholm A, Medina JF, Radmark O et al. Leukotriene A$_4$ hydrolase: abrogation of the peptidase activity by mutation of glutamic acid-296. Proc Natl Acad Sci USA 1992; 89:9141-45.

33. Carpenter FH, Vahl JM. Leucine aminopeptidase (bovine lens). mechanism of activation by Mg^{2+} and Mn^{2+} of the zinc metalloenzyme, amino acid composition, and sulfhydryl content. J Biol Chem 1973; 248:294-304.

34. Allen MP, Yamada AH, Carpenter FH. Kinetic parameters of metal-substituted leucine aminopeptidase from bovine lens. Biochem 1983; 22:3778-83.

35. Thompson GA, Carpenter FH. Leucine aminopeptidase (bovine lens). The relative binding of cobalt and zinc to leucine aminopeptidase and the effect of cobalt substitution on specific activity. J Biol Chem 1976; 251:1618-24.

36. Oettgen HC, Taylor A. Purification, preliminary characterization, and immunological comparison of hog lens leucine aminopeptidase (EC 3.4.11.1) with hog kidney and beef lens aminopeptidases. Anal Biochem 1985; 146:238-45.

37. Taylor A, Volz KW, Lipscomb WN et al. Leucine aminopeptidase from bovine lens and hog kidney. comparison using immunological techniques, electron microscopy, and x-ray diffraction. J Biol Chem 1984; 259:14757-61.

38. Shen C-C, Melius P. Leucine aminopeptidase from swine kidney: purification, molecular weight, subunit and amino acid composition. Prep Biochem 1977; 7:243-56.

39. Himmelhoch SR. Leucine aminopeptidase: a zinc metalloenzyme. Arch Biochem Biophys 1969; 134:597-602.

40. Smith EL, Spackman DH. Leucine aminopeptidase. V. Activation, specificity, and mechanism of action. J Biol Chem 1955;

212:271-99.

41. Lin W-Y, Van Wart HE. Hydrolysis of dansyl-peptide substrates by leucine aminopeptidase: origin of dansyl fluorescence changes during hydrolysis. Biochem 1988; 27:5054-61.

42. Lin W-Y, Lin SH, Van Wart HE. Steady-state kinetics of hydrolysis of dansyl-peptide substrates by leucine aminopeptidase. Biochem 1988; 27:5062-68.

42a. Taylor A, Tisdell, FE, Carpenter FH. Leucine aminopeptidase (bovine lens): synthesis and kinetic properties of ortho, meta, and para substituted leucyl-anilides. Arch Biochem Biophys 1981; 210:90-97.

43. Lin SH, Van Wart HE. Effect of cryosolvents and subzero temperatures on the hydrolysis of L-leucine-p-nitroanilide by porcine kidney leucine aminopeptidase. Biochem 1982; 21:5528-33.

44. Lin W-Y, Lin SH, Morris RJ et al. Stopped-flow cryoenzymological investigation of the presteady-state kinetics of hydrolysis of Leu-Gly-NHNH-Dns by leucine aminopeptidase. Biochem 1988; 27:5068-74.

45. Prescott JM, Wilkes SH, Wagner FW et al. *Aeromonas* aminopeptidase. improved isolation and some physical properties. J Biol Chem 1971; 246:1756-64.

46. Prescott JM, Wilkes SH. *Aeromonas* aminopeptidase. Methods Enzymol 1976; 45:530-43.

47. Prescott JM, Wagner FW, Holmquist B et al. Spectral and kinetic studies of metal-substituted *aeromonas* aminopeptidase: nonidentical, interacting metal-binding sites. Biochem 1985; 24:5350-56.

48. Bayliss ME, Prescott JM. Modified activity of *aeromonas* aminopeptidase: metal ion substitutions and role of substrates. Biochem 1986; 25:8113-17.

49. Ben-Bassat A, Bauer K, Chang S-Y et al. Processing of the initiation methionine from proteins: properties of the *escherichia coli* methionine aminopeptidase and its gene structure. J Bacteriol 1987; 169:751-57.

50. Chang Y-H, Teichert U, Smith JA. Purification and characterization of a methionine aminopeptidase from *saccharomyces cerevisiae*. J Biol Chem 1990; 265:19892-97.

51. Miller CG, Strauch KL, Kukral AM et al. N-terminal methionine-specific peptidase in *salmonella typhimurium*. Proc Natl Acad Sci USA 1987; 84:2718-2722.

52. Kendall RL, Bradshaw RA. Isolation and characterization of the methionine aminopeptidase from porcine liver responsible fro the cotranslational processing of proteins. J Biol Chem 1992; 267:20667-73.

53. Wallner BP, Hession C, Tizard R et al. Isolation of bovine kidney leucine aminopeptidase cDNA: comparison with the lens enzyme and tissue-specific expression of two mRNAs. Biochem 1993; 32:9296-9301.

METHIONINE AMINOPEPTIDASE: STRUCTURE AND FUNCTION

Ralph A. Bradshaw and Stuart M. Arfin

INTRODUCTION

Aminopeptidases (APs) that are capable of removing methionine residues from the N-terminal end of peptides and proteins can broadly be grouped into two classes: those which are nonspecific and generally act processively, i.e., they remove other residues as well as methionine and will continue to degrade the substrate until they encounter an unacceptable residue or sequence; and those which selectively remove methionine and are not processive. The former include such well-characterized enzymes as leucine aminopeptidase and aminopeptidase M and presumably participate in general degradation processes, while the latter are likely to be more highly regulated and have more narrowly delineated physiological roles.

The specific removal of methionine residues appears to be primarily associated with two processes: protein synthesis and in the maturation or degradation of specific bioactive peptides. The involvement of MetAPs in protein synthesis is primarily a cotranslational event in eukaryotes but appears also to be of importance as a post-translational event in prokaryotes (preceded by the removal of the N-formyl group).[1] There are a few documented cases for the post-translational removal of methionine in eukaryotes, particularly as associated with the processing of the cytoskeletal

Aminopeptidases, edited by Allen Taylor. © 1996 R.G. Landes Company.

protein actin, but these apparently require prior acetylation.[2] For the most part however, the removal of methionine is thought to occur when nascent chains reach 20-40 amino acids in length.[3] This has led to the general supposition that these enzymes are ribosomally associated. However, specific evidence in support of this is largely lacking. MetAPs have also been reported to be associated with mitochondria and with the microsomal fractions.[4,5] The first may be related to prokaryotic MetAPs in view of the fact that the limited amount of protein synthesis carried out in the mitochondrion appears to be initiated with N-formyl methionine. The latter group may be related to processing of bioactive peptide precursors or to their degradation. However, little detailed information is available for the enzymes involved in these pathways.

Most MetAPs identified to date are metalloenzymes which use primarily zinc or cobalt ions as cofactors.[6] The zinc-dependent APs are widely distributed in both prokaryotes and eukaryotes and generally display broad specificity. The cobalt-containing APs are also well characterized and apparently provide the enzymes responsible for processing the amino terminal residue during or immediately after translation. Both sequence and structural information is available for these enzymes, which can be broadly subdivided into two distantly related families.[7] The structural and functional properties of the cobalt-dependent MetAPs is described below.

COTRANSLATIONAL PROCESSING
OF EUKARYOTIC PROTEINS

It has been appreciated for over 25 years that the principal amino acid initiating protein synthesis in eukaryotic organisms is methionine.[3] Only a very small number of proteins have been found to be initiated with codons other than ATG (and they usually code for leucine residues). However, it has also been appreciated for an even longer period that mature proteins have a variety of N-termini, and, in fact, only a relatively small number appear to have an amino terminal methionine.[8] It has also been appreciated since the late 1950s that a considerable number of eukaryotic proteins, particularly those that occur intracellularly, contain blocking groups covalently attached to the α-amino group of the protein.[9] An acetyl moiety is most common, although several other types of blocking groups have also been identified.[10] The acetyl group is transferred (from acetyl-CoA) also apparently primarily as a cotranslational process. The clear absence of methionine from the N-terminal position of such a large percentage of proteins (including those with N-acetyl groups) clearly indicates the presence of a MetAP-type enzyme capable of processing polypeptide chains.

The first indications of the selectivity of this enzyme were derived from analyses of the protein sequence data base[8,11,12] and the study of yeast isocytochrome C mutants.[13] Initial studies suggested that the methionine residue was retained when polar or larger residues were in the penultimate position, whereas that was not the case when smaller hydrophilic residues occupied this position. Eventually, as correctly presaged by Sherman and colleagues,[13] it was found that the size (as opposed to the character) of the residue adjacent to the methionine appeared to be the primary specifying influence.[14,15] This was confirmed using site-directed mutagenesis in several studies in both yeast and higher eukaryotes. These studies indicated that, with very few exceptions, proteins in which the penultimate position is occupied by glycine, alanine, serine, threonine, valine, cysteine or proline lose their initiator methionine, whereas the remaining 13 amino acids prevent the removal of this residue. These studies also showed that glycine, alanine, serine and threonine were generally N-acetylated (after the removal of methionine), whereas proteins in which the penultimate residues were aspartic acid, glutamic acid or asparagine (which cause the retention of the methionine) were also acetylated. Thus, the 20 amino acids that can occupy the penultimate position following the initiator methionine can direct the action (or inaction) of the enzymes MetAP and N-acetyltransferase in the formation of four classes of proteins with respect to their N-terminal structures (Fig. 4.1). It should be emphasized that this information is lost in proteins that undergo more substantive proteolytic processing such as is found for extracellular proteins or proteins transported into the mitochondria where substantive peptides (approximately 20 amino acids) are removed from the N-terminus. The well-defined specificity observed in cotranslational processing, particularly with respect to the cleavage of Met-Pro sequences, has been an important identifying characteristic of the cobalt-containing MetAPs.

PROKARYOTIC AND YEAST MetAP

The first MetAPs to be defined chemically and sequentially were those derived from the prokaryotes *E. coli*[16] and *Salmonella*

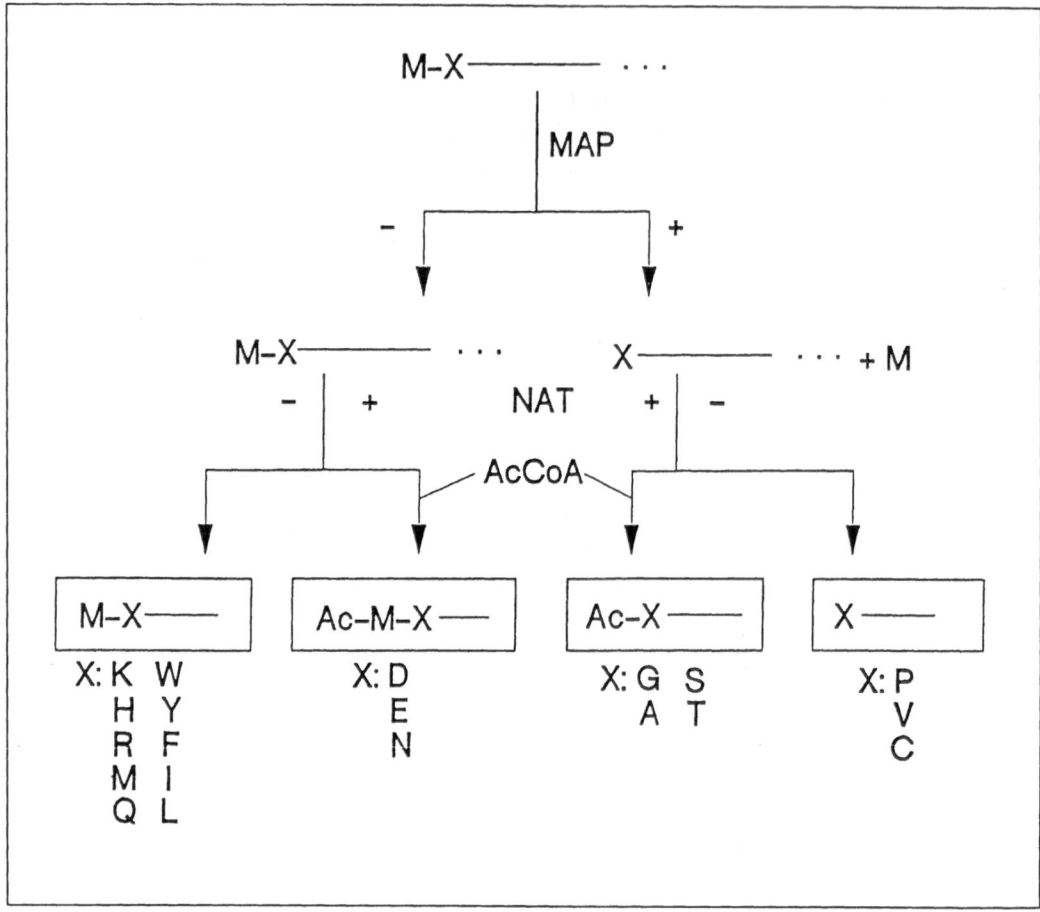

Fig. 4.1. Proposed pathway for the cotranslational modification of eukaryotic proteins by methionine aminopeptidase (MAP) and Nα-acetyl transferase (NAT). X represents, the penultimate residue. Dots indicate nascent polypeptide chains; (+) and (–) indicate positive or negative actions of the enzymes. Amino acid one letter code: A, Ala; C, Cys; D, Asp; E, Glu; F, Phe; G, Gly; H, His; I, Ile; K, Lys; L, Leu; M, Met; N, Asn; P, Pro; Q, Gln; R, Arg; S, Ser; T, Thr; V, Val; W, Trp; Y, Tyr. Reprinted with permission from Bradshaw RA, Trends Biochem Sci 1989; 14: 276-279.

typhimurium.[17] These enzymes, of approximately 30 kDa, contain cobalt ion, and cleave peptides with a specificity entirely in concert with that described above for the cotranslational processing of proteins in eukaryotes. However, the process of methionine excision in prokaryotes is clearly a post-translational event (and must be coupled to a deformylase enzyme). It was therefore of interest when the MetAP of yeast (*S. cerevisae*) was isolated, characterized and cloned.[18,19] This enzyme is clearly close-related to the prokaryotic enzymes but is somewhat larger in molecular weight (approximately 43 kDa). The cobalt-containing domain of the prokaryotic enzymes is clearly delineated in the C-terminal portion of the yeast enzyme, and the two enzymes show approximately 40% identity. The cobalt-binding ligands (identified from the x-ray crystallographic analysis of the *E. coli* enzyme)[20] are retained in the yeast enzyme. The amino-terminal extension that distinguishes the yeast MetAP from the prokaryotic forms contains some 125 residues and is characterized by a putative zinc finger region in the amino-terminal end. The whole enzyme does bind zinc, and truncated versions of this enzyme, lacking this region, bind only cobalt ions.[21] Clearly the zinc present in the mature enzyme is bound in this region and is therefore presumably not directly related to catalytic events. It has been suggested that this domain may be important in the association of yeast MetAP to ribosomes, presumably through interaction with nucleic acid moieties of that organelle. Interestingly, the amino-terminal domain containing the zinc finger can be removed from the intact yeast MetAP by trypsin indicating that the two domains are presumably independently folded and form distinct entities. A recombinant derivative which lacks the same region was significantly less effective in rescuing the slow growth phenotype of a *map I* mutant, which provides additional evidence for the importance of the zinc finger region. The recombinant protein lacking the zinc finger did not differ significantly in catalytic properties from the native enzyme when assayed in vitro.

The amino terminal 10 residues of yeast MetAP may also be involved in physiological processes.[19] The mature protein isolated from yeast lacks this 10 residue sequence, and its removal may be of some importance in downstream physiological events, particularly involving the zinc domain which lies immediately adjacent to this sequence. Importantly, the deletion of the gene for this protein from yeast is not lethal, and it has been suggested that there may be alternative N-terminal processing pathways in this organism.

The 3-dimensional structure of the *E. coli* MetAP has been determined by x-ray analysis to 2.4 Å resolution.[20] There is an internal pseudo 2-fold symmetry which structurally relates the first and second halves of the molecule (Fig. 4.2). Each of the two halves is characterized by an antiparallel β-pleated sheet flanked by two helical segments and a C-terminal loop. The interface in the center of the molecule is provided by interactions between the

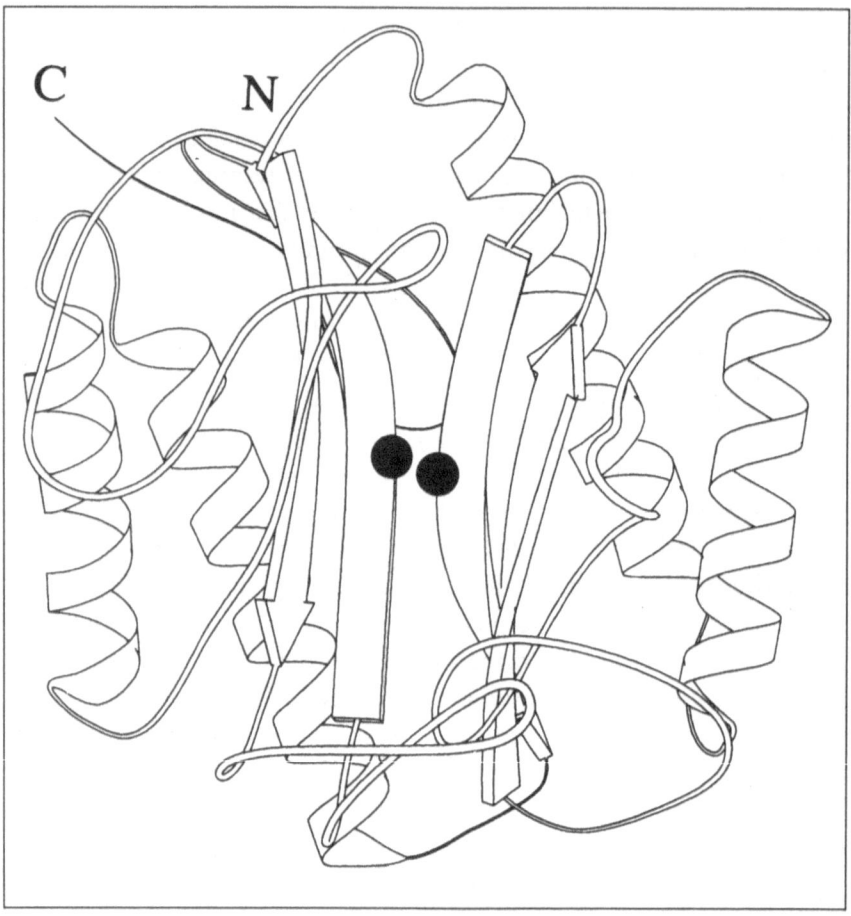

Fig. 4.2. Ribbon diagram showing the "pita-bread" fold of E. coli MetAP. The view direction is essentially parallel to the local 2-fold axis of symmetry and the active site is marked by the two cobalt ions shown as solid circles. Reprinted with permission from Bazan JF et al, Proc Natl Acad 1994; 91:2473-2477.

β-pleated sheet segments. The two cobalt atoms are closely juxta-posed in the β-pleated sheets and are liganded by the side chains of residues provided by aspartic, glutamic and histidine residues (97, 108, 204, 235 and 171) (see Fig. 4.3). The ligands are bound in approximately octahedral coordination. Of particular interest is the fact that the fold defined in this enzyme is unique and has not been observed previously in other proteins. Thus, the cobalt binding domain of this family appears to represent a new class of proteolytic enzymes, albeit that there have already been identified

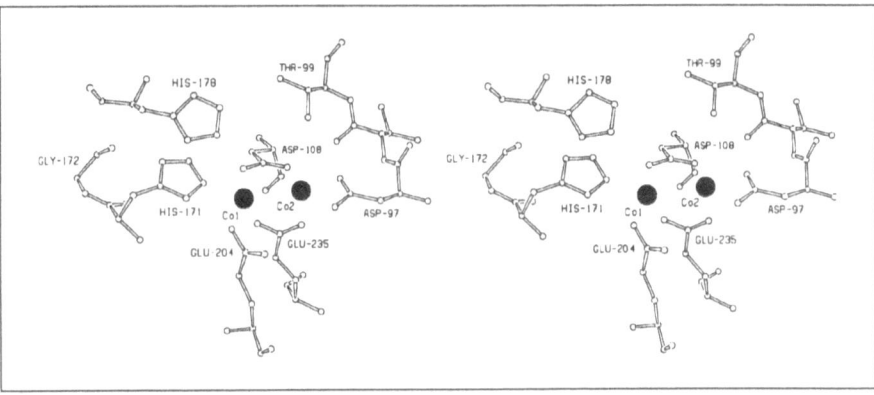

Fig. 4.3. Stereoview of the active site of E. coli MetAP including cobalt ions and protein ligands. Reprinted with permission from Roderick SL et al, Biochemistry 1993; 32:3907-3912, copyright 1993 American Chemical Society.

a number of additional proteins of generally related specificity that will certainly be similar in structure (see below).

HIGHER EUKARYOTIC METAPs

The first isolation and characterization of a higher eukaryotic MetAP was reported by Kendall and Bradshaw.[22] This protein, isolated to homogeneity from pig liver, also contained cobalt co-factors and had the same catalytic profile associated with the cotranslational processing steps. However, it was determined to be somewhat larger, showing an apparent molecular mass of 67,000 Da on SDS-PAGE electrophoresis. The catalytic activity of the enzyme was not affected by zinc ions, but the zinc content of the protein was not determined independently.

Partial amino acid sequence analysis of isolated fragments provided the means for obtaining cDNA clones from both porcine and human libraries.[7] Neither clone was full-length, and 5' RACE (rapid amplification of cDNA ends) technology was required to obtain the complete human sequence. This clone contained 2,569 nt with a single open reading frame that predicted a corresponding protein sequence of 478 amino acids (calculated molecular weight 52,832 Da). The substantial discrepancy between the calculated and the observed molecular weights suggests that either substantive post-ribosomal covalent modification occurs or that the observed molecular weight is anomalously high. In this regard, the amino-terminal sequence contains several long stretches of highly polar (acidic and basic) residues, and the occurrence of similar structures in other proteins has been shown to also produce anomalous molecular weight shifts of this magnitude on SDS-PAGE. Although post-ribosomal modifications cannot yet be ruled out, it seems likely that the true molecular weight is probably much closer to the calculated molecular weight.

Amino acid sequence alignment of the carboxyl terminal portion of the human (and the porcine) cDNA clones revealed a limited but statistically significant match to the cobalt-binding regions of the prokaryotic and yeast MetAPs described above (Fig. 4.4). Importantly, the ligands defined in the 3-dimensional structure of the *E. coli* enzyme[20] were also retained in these alignments (see Fig. 4.4). Model building experiments also established that the human sequence could be accommodated in the *E. coli* (in the catalytic domain region) without distortion of the main chain backbone of that enzyme, despite the presence of extensive deletions and additions characterizing the two comparisons. In each case, these segments occur in potential surface loop structures and therefore do not likely affect the principal organization of the backbone. Thus, based on the enzyme specificity, the common utilization of the cobalt cofactor and the limited sequence homology, it can be concluded that the porcine and human MetAPs are indeed related to and are likely to be homologous with the prokaryotic and yeast enzymes. It should be pointed out that this potential relationship was identified on the basis of computer sequence comparisons in the absence of any information that the amino acid sequence corresponding to human MetAP was, in fact, a metallo AP (see below).[23]

Fig. 4.4 (opposite). *Primary sequence alignment of E. coli MAP (E.C.), S. cerevisiae MAP 1 (S.c. I), human MAP (H.s.), S. cerevisiae hypothetical protein YBL091c (S.c. II) and a partial open reading frame from* Methanothermus fervidus *(M.f.). Blank spaces indicate that the amino acid residues are identical to those in the human sequence; dashes indicate deletions introduced to maximize the alignments. The three dots at the end of the M. fervidus sequence indicates that it is derived from a partial clone and that the sequence extends beyond this point for an unknown distance. Lower case letters denote conservative substitutions. Residues that are cobalt ligands in the structure of E. coli MAP and are conserved in the other sequences are boxed. Regions of substantial homology between human MAP and one or more of the other sequences are indicated by stippling. Reprinted with permission from Arfin SM et al, Proc Natl Acad Sci USA 1995, in press.*

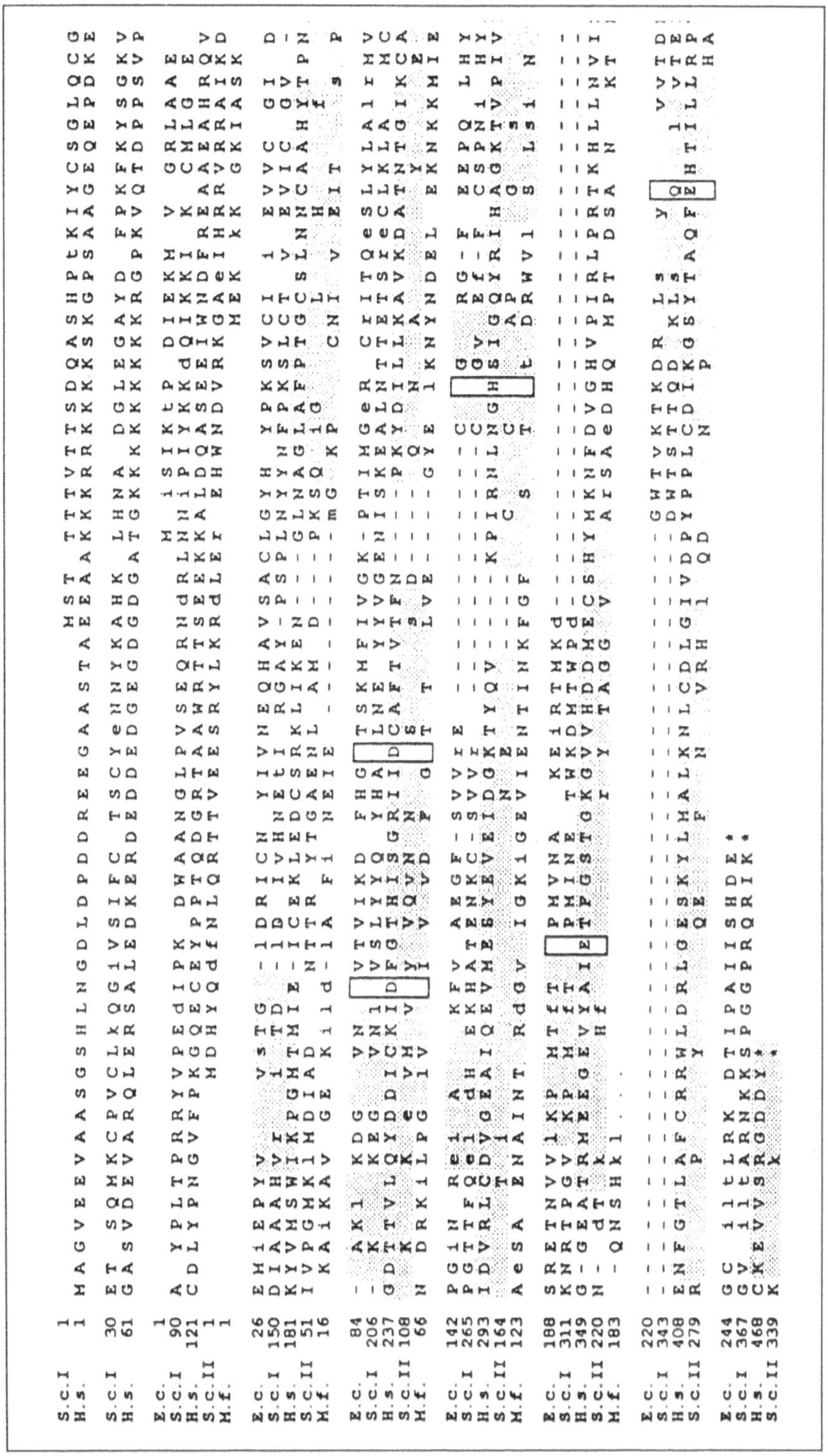

RELATIONSHIP OF HUMAN MetAP
TO OTHER PROTEINS

One of the more interesting observations following the purification and characterization of porcine MetAP (and subsequently the corresponding human protein) was that, based on the partial sequences obtained from isolated fragments, porcine MetAP was indeed similar (if not identical) to a previously reported protein termed p67, a protein associated with the eIF2α eukaryotic initiation complex involved in protein synthesis.[24] p67 inhibited the phosphorylation of the eIF2α subunit of this complex, a key regulatory step in translation. This protein was cloned and sequenced from rat. The human sequence is highly similar except at each terminus. At the amino-terminal end there are a number of point mutations, suggestive of species variations, and at the C-terminal end, there is a frameshift difference that renders the carboxyl-terminal sequence of the human protein completely different from that reported for rat.[7] This likely arises from a technical problem rather than representing a true species variation. The human sequence is, interestingly, much more similar to the prokaryotic and yeast MetAPs in this region. It was, in fact, the rat p67 sequence that was identified by Bazan et al[23] as a putative metallo AP. The determination that this sequence in fact corresponded to that for the isolated protein (porcine MetAP) confirmed this proposal.

The relationship between rat p67 (functioning as a regulatory subunit in the initiation complex) and the MetAP activity is not yet clear. Rat p67 is reported to be heavily glycosylated,[25] which has been suggested to be the basis for the differences between the observed and calculated molecular weights (as also noted for porcine MAP; see above). It is interesting that a protein found to ultimately function on the nascent chains could, based on this information, be recruited in a stoichiometric fashion to the initiation complex on the ribosome. Thus, the molecule could conceivably function both as a regulatory protein and as a catalyst for modifying the growing polypeptide chain. Its recruitment to each initiation complex would guarantee the association with each functioning ribosome.

Human MetAP (p67) is clearly related to a larger family of enzymes which includes prolidase, aminopeptidase P and creatinase.[23] A partial sequence from *Methanothermus fervidus* was also observed to be similar to this group of enzymes, suggesting it represents a prokaryotic MetAP that is related to the human enzyme.[26]

TWO FAMILIES OF METAPS

As described above, the MetAPs isolated from *E. coli*,[16] *Salmonella typhimurium*,[17] *B. subtilis*[27] and *S. cerevisae*[19] are clearly homologous to those identified and cloned from porcine and human sources.[7] However, there are substantive differences that separate the first group from the higher eukaryotes, which raises the question as to whether the significant structural and sequence variations observed between the two represents a major evolutionary shift or whether they are representative of two subfamilies of cobalt-containing APs.

Strong evidence in support of the latter view has been forthcoming from the identification of an open reading frame (ORF)[27] in yeast which is significantly more closely related to the human MetAP cDNA than to the sequence identified as yeast MetAP. This ORF has ~80% homology to the human species in the putative cobalt-containing catalytic domain. Although it has not yet been demonstrated that the protein corresponding to this ORF is indeed a metallo AP, this appears highly probable. In addition, the presence of a similar structure highly related to human MetAP (and the yeast ORF) in *M. fervidus* establishes that this subgroup extends into the prokaryotes. Thus, it suggests the existence of two MetAP families which have been designated type I and type II (Fig. 4.5).[7] The type I enzyme corresponds to the earlier identified prokaryotic species, including the *E. coli* enzyme, and is defined by the fold and characteristics determined for that structure. The first MetAP identified for yeast is clearly also of the type I family. In contrast, the human and porcine enzymes and the yeast ORF (and probably the partial sequence of *M. fervidus*) represent a second family (designated type II). There is a homologous relationship between the two families, but they are probably separated by some considerable evolutionary distance. Features of the two families that have been retained included the catalytic specificity and mechanism (including the dependence on cobalt) and an apparently similar 3-dimensional fold in the catalytic domain. What clearly separates the two families are variable amino-terminal extensions and the presence of substantial extra sequences in the

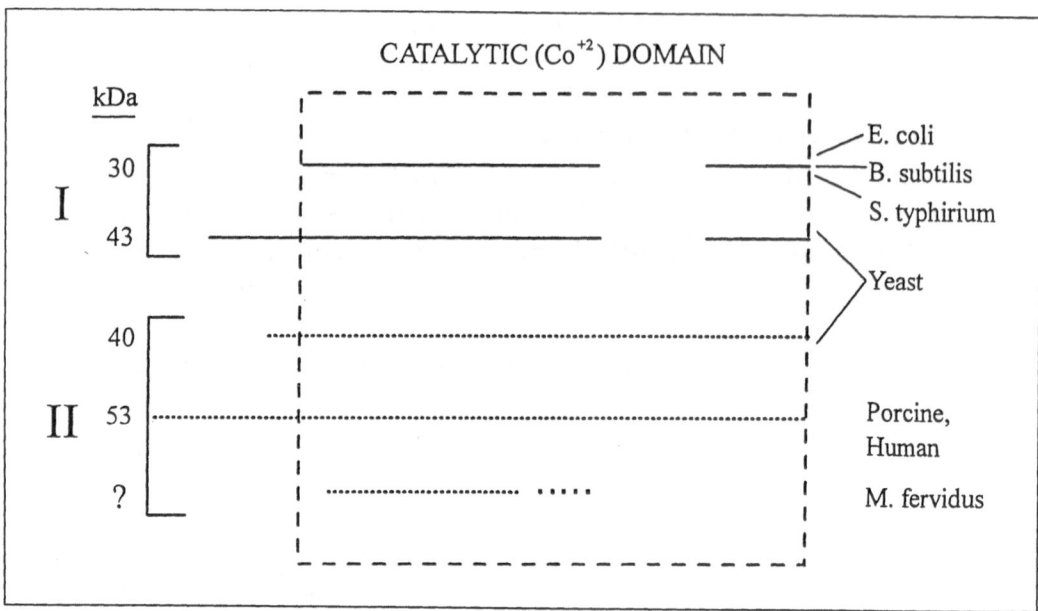

Fig. 4.5. Schematic representation of the two proposed subfamilies of MetAP. Reprinted with permission from Arfin SM et al, Proc Natl Acad Sci USA 1995, in press.

type II proteins. The prokaryotic form of the type I enzymes (by definition) lack such a domain, while yeast type I and human type II have considerable amino-terminal extensions, which are quite different from each other (Figs. 4.4, 4.5). The yeast type II (as represented by the ORF) has only a relatively short amino-terminal extension, which is unlike either the yeast type I or the human type II. It lacks the zinc finger of yeast type I and the long polar sequences of amino acids characteristic of the human type II. The putative *M. fervidus* protein also lacks this amino-terminal extension (assuming the remainder of the reading frame extends through and terminates at the end of the carboxyl-terminal domain as is found in all other MetAPs, studied to date).

The finding of two MetAP families in yeast (and probably in prokaryotes and higher eukaryotes as well) explains the lack of lethality of the deletion of the type I enzyme in yeast (if it is assumed that the type II enzyme can also function as a cotranslational processing enzyme). It apparently retains the correct specificity if one can extrapolate to the highly similar type II enzyme in human. It is, of course, possible that the two enzymes have different physiological functions. For example, one might function as a cotranslational processing enzyme, while the other may function downstream in various post-translational events. It is, after all, not rigorously proven that all amino-terminal processing of initiator methionine is indeed a truly cotranslational event.

SUMMARY AND FUTURE PROSPECTIVE

The MetAPs constituting the cobalt-dependent class (which apparently occur in two subfamilies) appear to be primarily involved in the processing of N-termini for protein synthesis.[28] The creation (along with N-acetyltransferases) of four classes of proteins has been related to protein turnover, particularly by the "N-end Rule"[29] (see chapter 1). In this pathway, proteolytic substrates are targetted for degradation following polyubiquitination, as directed by the ubiquitin ligase (E3) that recognizes amino-terminal amino acid residues of both the basic and bulky hydrophobic type.[30] There are ancillary (secondary and tertiary) pathways that allow the processing of acidic and amidated residues as well. The conservation of the specificities of MetAP can be taken as an argument for the importance of this pathway, although direct stability measurements suggest that the proposed specificity for stabilizing and destabilizing residues in the N-end Rule is only well matched to MetAP in yeast and shows serious deviations in higher organisms and prokaryotes.[31,32] The significance of these observations is not yet clear. The lack of identification of a significant number of protein substrates that are degraded by this pathway also raises questions regarding its physiological importance. The specificity of MetAP may, in fact, be more important in preventing premature degradation by this pathway than to act as a means for funneling proteins into it. Thus, a yeast null mutant in which both genes are eliminated (type I and type II) may still retain its viability as only the proteins in which methionine is removed would presumably be effected. Since these generally are the so-called "housekeeping" proteins that are characterized predominantly by N-terminal alanine and serine (which become subsequently N-acetylated) this may not become a serious problem. Only the proteins with penultimate residues of valine, cysteine and proline might be adversely affected, since, in some cases, these N-terminal residues are thought to be of key importance in the function of the protein.[28] The inability of these proteins to be processed could be the source of any deleterious effects observed from these mutations.

Future research will be required to confirm the proposed relationship between the type I and type II enzymes and to determine their relative physiological roles in intracellular metabolism. This information will also be valuable in ascertaining whether the cobalt-containing family of MetAPs function in a greater physiological spectrum of activities than in N-terminal processing during and immediately following translation.

REFERENCES

1. Arfin SM, Bradshaw RA. Mechanisms of regulated intracellular protein degradation. The Encyclopedia of Molecular Biology: Fundamentals and Applications. Weinham, Germany: VCH Verlags GmbH, 1995; in press.
2. Sheff DR, Rubenstein PA. Isolation and characterization of the rat liver actin N-acetylaminopeptidase. J Biol Chem 1992; 267: 20217-20224.
3. Jackson R, Hunter T. Role of methionine in the initiation of haemoglobin synthesis. Nature 1970; 227:672-676.
4. Braun HP, Schmitz UK. Purification and sequencing of cytochrome b from potato reveals methionine cleavage of a mitochondrially encoded protein. FEBS Ltrs 1993; 316:128-132.
5. Termignoni C, Freitas JO Jr, Guimarães JA. Methionyl aminopeptidase from rat liver: distribution of the membrane-bound subcellular enzyme. Molec and Cell Biochem 1991; 102:101-113.
6. Taylor A. Aminopeptidases: towards a mechanism of action. Trends Biochem Sci 1993; 18:167-172.
7. Arfin SM, Kendall RL, Hall L et al. Eukaryotic methionine aminopeptidases: Two classes of cobalt-dependent enzymes. Proc Natl Acad Sci USA 1995; in press.
8. Driessen HPC, de Jong WW, Tesser GI et al. The mechanism of N-terminal acetylation of proteins. CRC Crit Rev Biochem 1985; 18:281-325.
9. Narita K. Isolation of acetylpeptide from enzymic digests of TMV-protein. Biochim Biophys ACTA 1958; 28:184-191.
10. Bradshaw RA. Protein translocation and turnover in eukaryotic cells. Trends Biochem Sci 1989; 14:276-279.
11. Jornvall H. Acetylation of protein N-terminal amino groups structural observations alpha-amino acetylated proteins. J Theor Biol 1975; 55:1-12.
12. Persson B, Flinta C, von Heijne G et al. Structures of the N-terminally acetylated proteins. Eur J Biochem 1985; 152:523-527.
13. Sherman F, Stewart JW, Tsunasawa S. Methionine or not methionine at the beginning of a protein. BioEssays 1985; 3:27-31.
14. Huang S, Elliott RC, Lui P-S et al. Specificity of cotranslational amino-terminal processing of proteins in yeast. Biochemistry 1987;

26:8242-8246.

15. Boissel J-P, Kasper TJ, Bunn HF. Cotranslational amino-terminal processing of cytosolic proteins. Cell-free expression of site-directed mutants of human hemoglobin. J Biol Chem 1988; 263:8443-8449.

16. Ben-Bassat A, Bauer K, Chang, S-Y et al. Processing of the initiation methionine from proteins: Properties of the *Escherichia coli* methionine aminopeptidase and its structure. J Bacteriol 1987; 169:751-757.

17. Miller CG, Strauch KL, Kukral AM et al. N-terminal methionine-specific peptidase in *Salmonella typhimurium*. Proc Natl Acad Sci USA 1987; 84:2718-2722.

18. Chang Y-H, Teicher U, Smith JA. Purification and characterization of a methionine aminopeptidase from *Saccharomyces cerevisiae*. J Biol Chem 1990; 265:19892-19897.

19. Chang Y-H, Teichert U, Smith JA. Molecular cloning, sequencing, deletion, and overexpression of a methionine aminopeptidase gene from *Saccharomyces cerevisiae*. J Biol Chem 1992; 267:8007-8011.

20. Roderick SL, Matthews BW. Structure of the cobalt-dependent methionine aminopeptidase from Escherichia coli: A new type of proteolytic enzyme. Biochem 1993; 32:3907-3912.

21. Zuo S, Guo Q, Ling C et al. Evidence that two zinc fingers in the methionine aminopeptidase from *Saccharomyces cerevisiae* are important for normal growth. Mol Gen Genet 1995; 246:247-253.

22. Kendall RL, Bradshaw RA. Isolation and characterization of the methionine aminopeptidase from porcine liver responsible for the cotranslational processing of proteins. J Biol Chem 1992; 267:20667-20673.

23. Bazan JF, Weaver LH, Roderick SL et al. Sequence and structure comparison suggest that methionine aminopeptidase, prolidase, aminopeptidase P, and creatinase share a common fold. Proc Natl Acad 1994; 91:2473-2477.

24. Wu S, Gupta S, Chatterjee N et al. Cloning and characterization of complementary DNA encoding the eukaryotic initiation factor 2-associated 67 kDa protein (p[67]). J Biol Chem 1993; 268: 10796-10801.

25. Datta B, Ray MK, Chakrabarti D et al. Glycosylation of eukaryotic peptide chain initiation factor 2 (eIF-2)-associated 67 kDa polypeptide (p67) and its possible role in the inhibition of eIF-2 kinase-catalyzed phosphorylation of the eIF-2 α-subunit. J Biol Chem 1989; 264:20620-20624.

26. Haas ES, Daniels CJ, Reeve JN. Genes encoding 5s rRNA and tRNAs in the extremely thermophilic archaebacterium *Methanothermus fervidus*. Gene 1989; 77:253-263.

27. Feldmann H, Aigle M, Aljinovic G et al. Complete DNA sequence of yeast chromosome II. EMBO J 1994; 13:5795-5809.

28. Arfin SM, Bradshaw RA. Cotranslational processing and protein

turnover in eukaryotic cells. Biochemistry 1988; 27:7979-7984.

29. Bachmair A, Finley D, Varshavsky A. In vivo half life of a protein is a function of its amino-terminal residue. Science 1986; 234:179-186.

30. Hershko A, Crechanover A. The ubiquitin system for protein degradation. Ann Rev Biochem 1992; 61:761-807.

31. Gonda DK, Bachmair A, Wunning I et al. Universality and structure of the N-end rule. J Biol Chem 1989; 264:16700-16712.

32. Tobias JW, Varshavsky A. Cloning and functional analysis of the ubiquitin-specific protease gene UBP1 of *Saccharomyces cerevisiae*. J Biol Chem 1991; 266:12021-12028.

GENETIC AND BIOCHEMICAL ANALYSIS OF YEAST AMINOPEPTIDASES

Yie-Hwa Chang

Genetic and biochemical analysis of yeast aminopeptidases dras tically increased our understanding of the physiological roles of aminopeptidases. A search of GenBank revealed that over 60 aminopeptidase genes have been cloned. Among them, seven are from *Saccharomyces cerevisiae*. In this chapter, I will concentrate on the molecular genetic and biochemical studies of these seven yeast aminopeptidases. They include three cytosolic nonspecific aminopeptidases, two vacuolar aminopeptidases, and two methionine aminopeptidases. For extensive coverage of other yeast proteases and their properties, please see references 1-3.

CYTOSOLIC NONSPECIFIC AMINOPEPTIDASES

The major nonlysosomal proteolytic system consists of a multifunctional proteinase, which shows a chymotrypsin-like, a trypsin-like and a peptidylglutamyl-peptide hydrolyzing activity in vitro.[4,5] The yeast homolog of this multifunctional proteinase complex, also named proteasome, is proteinase yscE.[6,7] Since this complex only processes proteins to peptide levels, aminopeptidases with broad specificity must play critical roles in downstream processing of the proteasome-derived peptides to amino acids. These free amino acids, after being charged with tRNA, can then be recycled for protein

Aminopeptidases, edited by Allen Taylor. © 1996 R.G. Landes Company.

synthesis. Three aminopeptidase genes, whose products are found in yeast cytosol, have been cloned and sequenced. They are *BLH1*, *AAP1*, and *APE2*.[8-10]

BLEOMYCIN HYDROLASE-1 (BLH1)

A mammalian aminopeptidase capable of cleaving bleomycin has been characterized.[11] Bleomycin is a glycopeptide which acts as an antitumor antibiotic in the treatment of human cancers. The antitumor function of bleomycin is limited by the activity of bleomycin hydrolase. Only malignant tissues showing low levels of bleomycin hydrolase are sensitive to bleomycin.[11]

On the basis of the deficiency to cleave the substrate Cbz-Leu-Leu-Glu-βNA at the α-naphthylamide bond in vitro, *pre3-2* mutant strains affected in the peptidylglutamyl-peptide hydrolyzing activity of the yeast proteasome had been isolated and characterized.[12] Cloning of the *BLH1* gene by Enenkel and Wolf was achieved by an in situ test measuring the restoration to wild type levels of the Cbz-Leu-Leu-Glu-βNA hydrolyzing activity in the *pre3-2* mutant strain C13-9C, which was transformed with a high copy yeast genomic library.[8]

The *BLH1* gene contains an open reading frame of 1449 nucleotides. It is mapped to the left arm of chromosome XIV and is located 500 base pairs upstream of the open reading frame of the *KEX2* gene.[13] The open reading frame codes for a protein of 483 amino acids with a predicted molecular mass of 55.4 kDa. The predicted amino acid sequence of BLH1 shares 48% identity with a 277-amino acid sequence fragment of rabbit bleomycin hydrolase.[14] Both yeast and rabbit bleomycin hydrolase showed significant homology to a 15-amino acid segment within the active site cysteinyl region of cysteine proteinase, which is located in the amino-terminal part of the protein.[14]

The purified yeast BLH1 can hydrolyze aminopeptidase substrates with a neutral, basic, or acidic amino acid residue. However, the peptide derivative Cbz-Leu-Leu-Glu-βNA, which is specifically cleaved at the β-naphthylamide bond by the peptidylglutamyl-peptide hydrolyzing activity of the proteasome, was not cleaved by bleomycin hydrolase. Enenkel and Wolf proposed the existence of an endopeptidase that can cleave Cbz-Leu-Leu-Glu-βNA and generates an unblocked aminoacyl derivative. This unblocked aminoacyl derivative is then hydrolyzed by bleomycin hydrolase. Only overexpression of bleomycin hydrolase by the cloned *BLH1* gene can yield detectable amounts of chromophore released from the aminoacyl derivative and thereby leads to the phenotype in vitro suppression of the *pre3-2* mutation.[8]

Biochemical and genetic data indicate that *BLH1* and *LAP3*, which code for aminopeptidase yscIII, are identical.[15] Deletion of the *BLH1* gene in wild type cells leading to a nonfunctional bleomycin hydrolase did not affect the peptidylglutamyl-peptide hydrolyzing activity of the proteasome, confirming that *pre3-2* and *blh1::URA3* are alleles of two different genes. Although deletion of the *BLH1* gene showed no consequence for cell growth, when *blh1* null mutants were exposed to bleomycin, the deficiency of bleomycin hydrolase led to lethality, indicating that the enzyme is able to protect cells from bleomycin-induced intoxication.

Since bacteria-like streptomyces can secrete glycopeptide antibiotics such as bleomycin, the protective function of BLH1 may be required for the survival of yeast cells grown in a natural environment. The yeast *blh1* mutant may be used as a tool to test the drug resistance and therapeutic application of antibiotics derived from the bleomycin group.[8]

AMINOPEPTIDASE YSCII

Leucine aminopeptidase mutations *lap1*, *lap2*, *lap3* and *lap4* were obtained through the loss of aminopeptidase activity on leu-β-naphthylamide in successive rounds of mutagenesis by Trumbly and Bradley.[15] Aminopeptidase yscII (also called LAP I) accounts for most of the cellular aminopeptidase activity in exponentially growing cells.[16] In commercial brewer's yeast, about half of the enzyme is found in the cytosol, the rest in the periplasmic compartment.

Yeast mutants lacking this activity have been isolated and characterized. Analysis of a mutant with a thermosensitive AP yscII activity, as well as gene dosage studies revealed that *APE2* is the structural gene of the enzyme.

The *APE2* sequence predicts a protein containing a polypeptide core of 97,368 Da with two potential N-linked glycosylation sites. The predicted translational product of the *APE2* gene lacks an amino-terminal hydrophobic segment which could function as a signal peptide. Furthermore, its amino acid sequence shows significant similarity to mouse BP-1/6C3 antigen, rat microsomal

```
       Yeast AAP1    21 ATTQMEATDARRAFPCFDEPNLKATFAVTL   50
       Yeast APII   132 ------P---------------------S--I--  161
   I   Mouse BP1    225 -A-DHEP----KS---------K-S-YSISI   254
       Human APN    208 -----Q-A---KS-------AM--E-NI--   237
       Rat   APN    204 -----Q-A---KS-------AM--S-NI--   233

       Yeast AAP1    76 TTFNTTPKMSTYLVA   90
       Yeast APII   187 -L-------------  201
  II   Mouse BP1    284 ---VKSVP------C  298
       Human APN    268 -E-H---------L-  282
       Rat   APN    262 -E-HP--------L-  276

       Yeast AAP1   155 FSAGAMENWGLVTYR  169
       Yeast APII   266 --------------  280
 III   Mouse BP1    366 -GT-----------  380
       Human APN    349 -N------------  363
       Rat   APN    344 -N------------  358

       Yeast AAP1   186 QRVAEVIQHELAHQWFGNLVTMDWWEGLWLNEGFA  220
       Yeast APII   297 ------V---------------------------  331
  IV   Mouse BP1    397 ----S-VA---V------T------DD--------  431
       Human APN    380 E--VT--A-------------IE--ND--------  414
       Rat   APN    375 E--VT--A-------------VD--ND--------  409

       Yeast AAP1   275 FDAISYSKGSSLLRMIS  291
       Yeast APII   386 ---------A-------  402
   V   Mouse BP1    489 --G------A-I---LQ  505
       Human APN    472 ---------A-V---L-  488
       Rat   APN    467 --S-T----A-V---L-  483
```

Fig. 5.1. *Sequence identity between Aap1, aminopeptidase yscII and mammalian aminopeptidases. Five highly conserved regions are designated by Roman numerals. A dash indicates an identical residue. Mouse BP1 is a mouse pre-B cell antigen. APN stands for aminopeptidase N.*

aminopeptidase N, human aminopeptidase N, aminopeptidase N of *E. coli* and to human leukotriene A₄ hydrolase, all of which are members of the zinc dependent metallopeptidase gene family (Fig. 5.1).[18-22] The similarity is especially striking in the region containing a zinc binding motif (VVQHELAHQW; residues 302-311 of AP yscII).

Leucine auxotrophs carrying the *ape2-1* mutation carboxypeptidase-yscY- and carboxypeptidase-yscS-negative background had been found to show a reduced growth when the leucine-containing peptides Leu-Gly or Leu-Leu were used as supplements on MV medium instead of leucine.[19] Otherwise no phenotype differences between *ape2-1* mutants and wild type cells could be detected.

ALANINE/ARGININE AMINOPEPTIDASE-1 (AAP1)

The yeast *AAP1* gene, encoding an aminopeptidase, was isolated based on its ability to suppress the temperature-sensitive growth on nonfermentable carbon source of *spr5*, a stationary phase regulatory mutant.[9] AAP1 was physically mapped to chromosome VIII between *PUT2* and *CUP1*. Sequence analysis of the *AAP1* gene showed a 1581-nucleotide open reading frame encoding a polypepetide of 526 amino acids with an approximate molecular mass of 59 kDa.

DR Caprioglio et al found that the *AAP1* gene product shows a high degree of sequence identity to several mammalian aminopeptidases, including the mouse pre-Bcell antigen BP1 (39.5% identity over 505 amino acids), rat aminopeptidase N (40% over 460 amino acids), and human intestine aminopeptidase N (37.5% identity over 483 amino acids). Lower but significant levels of amino acid sequence identity were found between Aap1 protein and human leukotriene A_4 hydrolase (25.5% identity over 357 amino acids) and *E. coli* aminopeptidase N (25.1 identity over 383 amino acids).[18-22] Interestingly, I found that the *AAP1* gene product also exhibits 92% amino acid sequence identity with that of yeast aminopeptidase yscII, which was ignored by Caprioglio et al (Fig. 5.1).

Among these aminopeptidases, five highly conserved regions were observed. These regions are between 15 to 35 amino acids in length. In these regions, the sequence identity between Aap1 protein, aminopeptidase yscII, and mammalian aminopeptidases are very high, ranging from 53% (region I) to 93% (region III). The similarity in these regions between the predicted amino acid sequence of Aap1 and aminopeptidase yscII and the sequences of the mammalian aminopeptidase N approached 100% (Fig. 5.1). No functions have been suggested for regions I, II, or III. Region IV has a high sequence identity to the zinc binding domains of several aminopeptidases, and region V shows homology to other peptidases with a putative proton donor at tyrosine residue 280.[9]

Insertional inactivation of the *AAP1* gene resulted in a decrease in glycogen accumulation and the loss of the major band of arginine/alanine aminopeptidase activity. Strains carrying the *AAP1* gene on high copy plasmid show an increase in the major arginine/alanine aminopeptidase activity. Although the *AAP1* gene is not essential for viability, the Aap1 protein positively affects glycogen accumulation in yeast.

Since this is the first aminopeptidase implicated in the regulation of glycogen accumulation, Aap1 may be extremely valuable in understanding the relationships between signal transduction and proteolysis in yeast. In addition, the interesting phenotype of yeast *aap1* null strains may be a good system for studying the structure/function relationships among yeast Aap1, yeast aminopeptidase yscII and their mammalian homologs.

YEAST VACUOLAR AMINOPEPTIDASES

Vacuoles in yeast function similarly to lysosomes in higher eukaryotes. Two endoproteinases, PrA and PrB, two carboxypeptidases, CpY and CpS, one dipeptidyl aminopeptidase, and two aminopeptidases have been found in yeast vacuole.[1,2] Here I will focus on the genetic studies of two vacuolar aminopeptidases, aminopeptidase I and aminopeptidase Y.

Aminopeptidase I

Yeast vacuolar aminopeptidase I was first isolated and characterized as a high molecular weight protein by Johnson in 1941.[24] This enzyme was established to be a Zn^{2+} containing metalloexopeptidase. Mutations in the *Saccharomyces cerevisiae LAP4* gene, which have been generated by Trumbly and Bradley, eliminated the activity of aminopeptidase I, and the *LAP4* gene has been genetically mapped to the left arm of chromosome XI.[15] Yeast strains with mutations in *LAP4* displayed normal growth rates, indicating that aminopeptidase I does not play vital role in cell growth or its absence can be compensated by other aminopeptidases.

Chang and Smith first cloned the genomic DNA encoding the vacuolar aminopeptidase I from a yeast EMBL3A genomic library.[25] The DNA sequence encodes a precursor protein containing 514 amino acid residues. The "mature" protein, whose N-terminal sequence was confirmed by automated Edman degradation, consists of 469 amino acids. A 45-residue presequence contains positively and negatively charged as well as hydrophobic residues, and its N-terminal residues could be arrayed in an amphiphilic α-helix (Fig. 5.2). This presequence differs from the signal sequences which direct proteins across bacterial plasma membranes and endoplasmic reticulum or into mitochondria. This gene was also cloned later by R. Cueva et al by a genetic complementation method.[26]

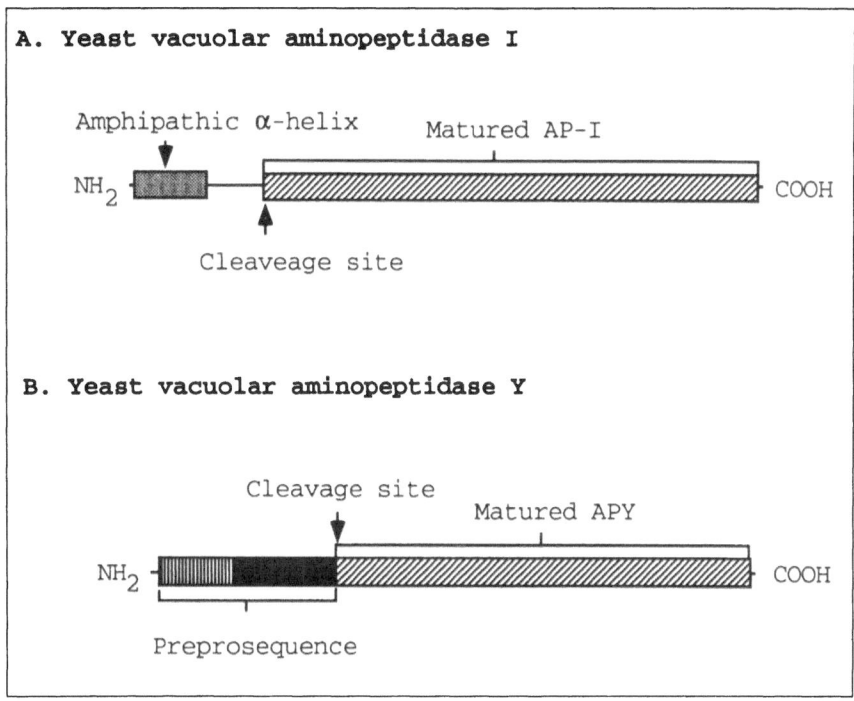

Fig. 5.2. Schematic presentation of yeast vacuolar aminopeptidase I and aminopeptidase Y. The cleavage site of API is between residues 45 and 46. The cleavage site of APY is between residues 56 and 57.

Aminopeptidase I was assumed to be glycosylated by Metz and Rohm.[27] However, later it was found by Klionsky et al that it is not glycosylated, and it does not enter the secretory pathway.[28] The API precursor remains in the cytoplasm and appears to translocate directly into the vacuole with a half-time of ~45 min. This is in contrast to the processing half-time ~6 min seen for carboxypeptidase Y, protease A or alkaline phosphatase.[29-31] The removal of the prosequence occurs in a *sec*-dependent manner. Neither the precursor nor the mature form of API are secreted into the extracellular fraction in *vps* mutants or upon overproduction, two additional characteristics of soluble vacuolar proteins that transit through the secretory pathway.[28] Overproduction of API results in both an increase in the half-time of processing and the stable accumulation of precursor protein, suggesting that API enters the vacuole by a post-translational process not used by most previously studied resident vacuolar proteins. A likely explanation for the lack of complete maturation upon overproduction, as proposed by Klionsky et al, is that one or more cytoplasmic components such as hsp 70 protein are required for maintaining the precursor in a translocation competent state. API will be a useful model protein to analyze this alternative mechanism of vacuolar localization.[32-35]

AMINOPEPTIDASE Y

Aminopeptidase Y is a another aminopeptidase localized in yeast vacuole. The enzyme is a single-chain molecule of 70 kDa, consisting of a 53 kDa protein and 17 kDa of sugar chains.[36] Aminopeptidase Y was inhibited by metal chelators and bestatin, like typical aminopeptidase. A unique feature of aminopeptidase Y was the effects of Co^{2+} on the activity. The hydrolytic activity of aminopeptidase Y toward amino acid-4-methylcoumaryl-7-amide (MCA) was greatly enhanced by Co^{2+}, whereas the NH_2-terminal amino acid-releasing activity from dipeptidyl-MCAs, tripeptides, and longer peptides was inhibited. Aminopeptidase Y was found as a 74 kDa zymogen in ABYS1 mutant cells in which the vacuolar proteases had been deleted. The zymogen is processed and activated to mature 70 kDa aminopeptidase Y by vacuolar protease B. Immunoadsorption of yeast vacuoles with anti-aminopeptidase Y antiserum resulted in loss of the major part of the aminopeptidase activities in this organelle.[36]

A yeast genomic DNA encoding aminopeptidase Y was isolated and mapped to chromosome II. It encodes 537 amino acids.[37] The "mature" protein, whose NH_2-terminal sequence was determined previously by analysis of the purified enzyme, consists of 481 amino acids, and the calculated molecular weight (52,900 Da) coincides with the value obtained by SDS-PAGE of the enzyme after removal of sugar chains, 53 kDa. The 56-residue preprosequence can be divided into two parts by putative processing sites for signal peptidase and conversion to the mature form. The 21-residue presequence has a hydrophobic stretch which may function as the signal sequence for transit through the endoplasmic reticulum, and the 35-residue prosequence accounts for the vacuolar proteases-deleted ABYS1 mutant and wild type mature enzyme.[37]

Deletion of aminopeptidase Y gene from *S. cerevisiae* showed no effect on cell growth. To estimate how much of the vacuolar aminopeptidase activities are due to aminopeptidase Y, Nishizawa et al compared aminopeptidase activities in yeast vacuoles of the disruption mutant with those of the wild type cells.[37] In the yeast mutant cells lacking aminopeptidase Y, vacuolar aminopeptidase activities toward Ala-4-MCA and Lys-MCA were 13 and 20% of the activity of the wild type. When Co^{2+} was added to the assay mixture, the remaining activity in the mutant vacuole was only 2.2 and 2.8% of that of the wild type, respectively.

To determine the physiological roles of these two vacuolar aminopeptidases, API and APY, it is necessary to knockout both aminopeptidase genes from *S. cerevisiae* and analyze its effect on cell growth under different conditions.

METHIONINE AMINOPEPTIDASES (METAPS)

Methionine aminopeptidases are responsible for the removal of the initiator-methionine from nascent polypeptides. Removal of the NH_2-methionine is essential for certain proteins to function normally in vivo. Examples for such proteins include: (a) protein kinases such as the catalytic subunit of cAMP-dependent protein kinase and p60[src]; (b) protein phosphatases such as calcineurin B; (c) proteins involved in transmembrane signaling such as the α-subunits of several G proteins; (d) proteins involved in the regulation of protein secretory vesicular trafficking through the Golgi stack such as Arf1p and Arf2p; (e) the endonuclease V in which the position of the αNH2 group affects its catalytic activity; and (f) the hemoglobins whose allosteric properties are strongly dependent on the N-terminal structure of their β subunits.[38-47]

METHIONINE AMINOPEPTIDASE-1

The first eukaryotic methionine aminopeptidase purified from *S. cerevisiae*, like the MetAPs from prokaryotes and porcine liver, is a cobalt-dependent metallopeptidase.[48-51] Favorable peptide substrates for MetAPs possess an NH_2-terminal methionine followed by a small and uncharged residue, which is in general agreement with the specificities of MetAPs predicted by the in vivo studies.[52-55] The x-ray structure of *E. coli* MetAP was recently determined to 2.4 Å resolution.[56] This bacterial MetAP contains two cobalt ions in the active site, and it appears to represent a new class of proteolytic enzymes.

The gene encoding MetAP1 from *S. cerevisiae* has been cloned and sequenced.[57] The open reading frame specifies a protein of

387 amino acids with a calculated MW of 43,269 Da. The N-terminal sequence matches perfectly with the deduced protein sequence beginning at residue 11. There are 10 additional amino acids (including 6 threonines and 1 serine) at the N-terminus. Recently, we found that these ten residues are not removed from yeast MetAP1 isolated from cells grown to mid-log phase. Mutant MetAP1 lacking these ten residues can fully rescue the slow-growth phenotype of the *map1* null strain (Guo and Chang, unpublished observation). However, we found that the mutant MetAP1 appears to be more stable than the wild type MetAP1 in cells grown to stationary phase. These findings led us to propose that these ten residues may play a role in MetAP1 turn-over, which is now under investigation in my laboratory.

The deduced amino acid sequence of the MetAP1 from *S. cerevisiae* is similar to those of the MetAPs from *E. coli* (42.7%), *S. typhimurium* (43.7%), and *B. subtilis* (40.4%), and there are four short but highly conserved regions (residues 158-166, 210-215, 286-297 and 319-333). Interestingly, the N-terminal domain contains two sequences that match the generalized consensus sequence associated with the zinc finger motif, $(Cys/His)-X_{2-4}-(Cys/His)-X_{2-15}-(Cys/His)-X_{2-4}-(Cys/His)$. The first one resembling a Cys_2-Cys_2 finger is located between residues 22-40, and the second one resembling a Cys_2-His_2 finger is located between residues 50-66. This unique combination of the zinc fingers is distinct from all other known zinc fingers. However, the individual zinc finger does show some interesting similarities with known zinc fingers. For example, the Cys_2-Cys_2 putative zinc finger of yeast MetAP1 shares significant similarity to the Cys_2-Cys_2 motif in the RING finger.[58] In both zinc fingers, there is a preference for: (1) Phe/Tyr before Cys_1; (2) a hydrophobic residue after Cys_2; (3) proline after Cys_3; and (4) a hydrophobic residue before Cys_4.[59] Comparative studies of the structure and function of these zinc fingers should reveal intriguing insights into the relationships between folding of zinc fingers and their binding specificities. The Cys_2-His_2 putative zinc finger of yeast MetAP1, on the other hand, also shares some notable similarities with those zinc fingers involved in RNA binding, even though they are not closely related to each other. These include a preference for Phe/Tyr before the Cys_1 and more than one aromatic residues between the Cys_2 and His_1 residues.

To determine whether yeast MetAP1 contains zinc fingers, we first asked whether yeast MetAP1 contains zinc in its N-terminal region. To address this question, a mutant MetAP1 lacking residues 2-69 was constructed, overexpressed, purified, and analyzed.[59] Metal ion analyses indicate that one mole of wild type yeast MetAP1 contains two moles of zinc ions and at least one mole of cobalt ion, whereas one mole of the truncated MetAP1 lacking the putative zinc fingers contains only a trace amount of zinc ions but still contains one mole of cobalt ion. These results suggest that those two zinc ions observed in the native yeast MetAP1 are located at the Cys/His rich region and the cobalt ion is located in the catalytic domain.[59] It is noteworthy that only one cobalt ion was detected in yeast MetAP1, while there are two cobalt ions in the bacterial MetAP1. There are at least three possible explanations for this discrepancy: (1) the cobalt content was underestimated due to extensive dialysis; (2) yeast MetAP1 contains only one cobalt ion; and (3) yeast MetAP1 contains only one tightly bound cobalt ion, and the loosely bound one was lost during dialysis.

The roles of the zinc fingers in catalysis was evaluated by comparing the k_{cat} and K_m values of the wild type MetAP1 and the truncated MetAP1 on peptide substrates, Met-Ala-Ser, Met-Gly-Met-Met, and Met-Leu-Phe. We found that deletion of the zinc fingers has little effect on the substrate specificity of yeast MetAP1. Thus, the zinc fingers of yeast MetAP1 do not appear to be involved in catalysis in vitro.[59]

To evaluate the effects of deletion of the zinc finger region on MetAP1 function in vivo, we asked how well the truncated MetAP1 could rescue the slow-growth phenotype of the yeast map1 null mutant. We found the doubling time of the yeast strain expressing mutant MetAP1(Δ2-69) (4 hrs) to be shorter than that of the yeast *map1* null mutant (6.5 hrs), but significantly longer than that of yeast strain containing wild type MetAP1 (2 hrs). We also found that the expression level of the truncated MetAP1(Δ2-69) was comparable to that of the wild type MetAP1 as indicated by enzyme assays. Thus, the slower than normal growth rate of the yeast cells expressing the truncated MetAP1 cannot be attributed to a deficiency of catalytically active MetAP1.

In addition, the roles of the cysteine and histidine residues in the zinc finger-like motifs were investigated by site-directed mutagenesis. The six cysteine residues in the putative zinc fingers were changed to serines and the two histidine residues to arginines. Yeast colonies expressing mutant MetAPs showed up 12 to 24 hours later than those expressing wild type MetAP1. Analysis of the doubling time of the yeast strains expressing different mutant MetAP1 revealed that yeast transformants expressing MetAP1 with single mutations have doubling times ranging from four to six hours, suggesting that these cysteine and histidine residues play important roles in MetAP1 function in vivo. Enzyme assays indicated that the expression level of each mutant MetAP1 is comparable to that of wild type MetAP1.[59]

It has been demonstrated that MetAP1 functions cotranslationally, but it has a relatively low affinity to its peptide substrates ($K_m \approx 0.2$ to 6 mM). Recently, we obtained evidence that yeast MetAP1 associates with ribosomes via its zinc fingers. Thus, we proposed that the zinc fingers may act as a binding domain linking MetAP1 to ribosomes such that the active site of yeast MetAP1 is located at a site near the exit of nascent polypeptides, thereby facilitating the processing function of wild type MetAP1 in cells. A mutant MetAP1 lacking functional zinc fingers may be either dissociated from ribosomes or associated with ribosomes in a different way such that the active site is away from the nascent polypeptides, which in turn reduces its efficiency in processing the initiator methionine in vivo, even though the in vitro enzyme assays indicate that deletion of the zinc fingers show no significant effects on MetAP1 activity.

METHIONINE AMINOPEPTIDASE 2

As described above disruption of the *MAP1* gene from *S. cerevisiae* is not lethal. However, since it has been shown that myristoylation is essential for cell growth and that N-myristoyltransferase requires a free NH_2-terminal glycine, it was expected that the retention of NH_2-methionine would block the myristoylation of certain proteins and in turn cause lethality. The only explanation for the finding that *map1* null strains are viable is that partial removal of the NH_2-terminal methionine occurs in the absence of MetAP1. Hence, it is very likely that there exist other enzyme(s) that can remove the NH_2-methionine in vivo. Recently, we cloned a gene from *S. cerevisiae* that can suppress the slow-growth phenotype of the *map1* null strain.[60] Sequence analysis

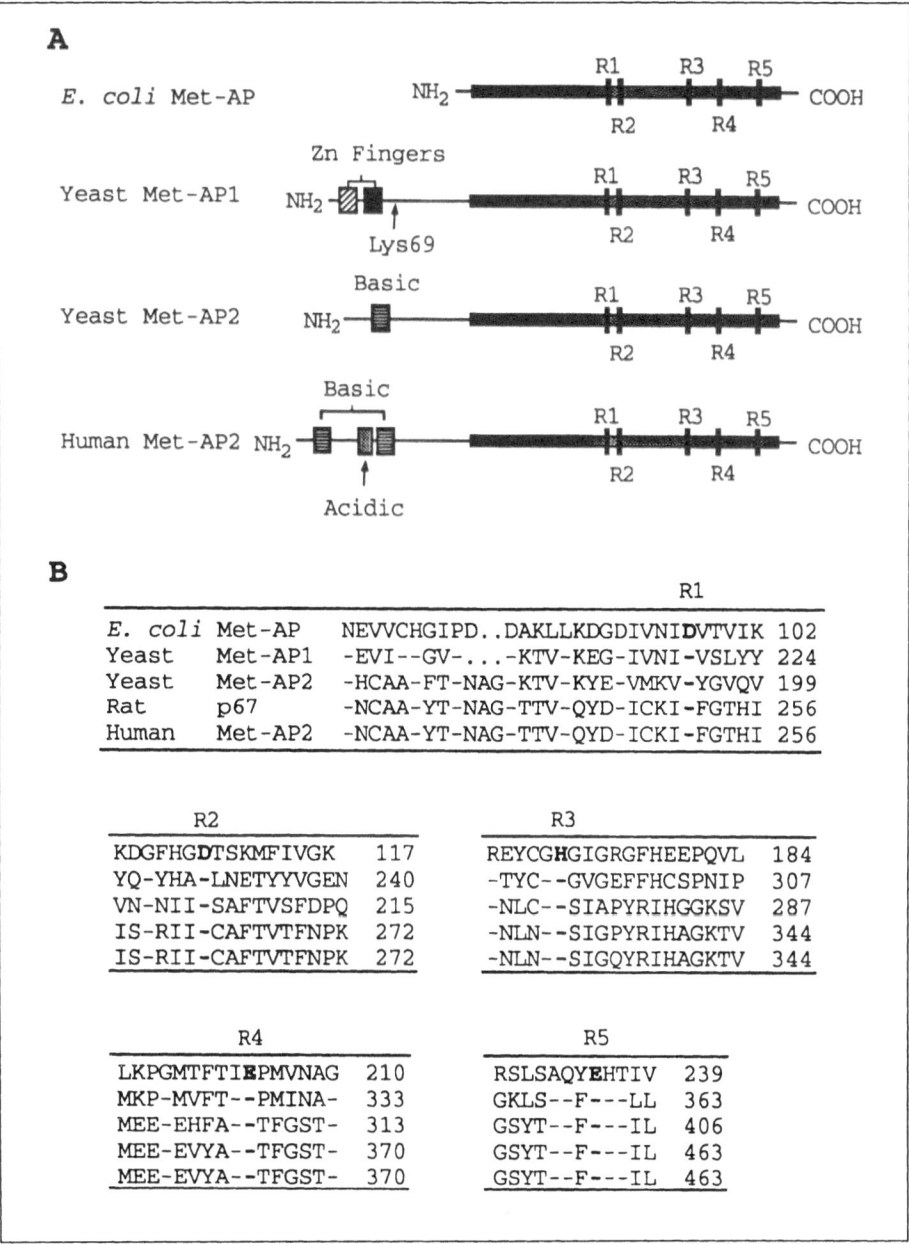

Fig. 5.3. Structural organization of E. coli, S. cerevisiae, and human MetAPs. (A) Schematic presentation of bacterial MetAP, yeast MetAP1, yeast MetAP2, and human MetAP2. R1-R5 indicate five residues which are conserved in all three MetAPs. (B) Amino sequences surrounding the five cobalt binding residues (bold-faced) conserved in bacterial, yeast, and human MetAPs as well as from rat p67. A dash indicates an identical amino acid, and a dot indicates that a gap was introduced to give optimal alignments.

revealed a single open reading frame encoding a protein of 421 amino acids. The deduced amino acid sequence of the suppressor gene shares a significant 22% identity with that of yeast MetAP1. All five residues involved in cobalt binding in *E. coli* MetAP were conserved among this suppressor gene product and MetAPs (Fig. 5.3).

To determine whether the suppressor gene encodes a methionine aminopeptidase, we overexpressed the suppressor gene product in the *map1* null background and asked whether the MetAP activity in this strain increased correspondingly. Compared to *map1* cells expressing a normal level of the suppressor gene product, the *map1* cells overexpressing the suppressor gene showed a 4- to 12-fold increase in MetAP activity, suggesting that this suppressor gene does encode an enzyme that has MetAP specificity activity. To obtain a second line of evidence, we used the epitope-tagging/immunoprecipitation approach and carried out MetAP assays. The hemagglutinin epitope was tethered to the N-terminus of the suppressor gene product. The immunoaffinity-purified tagged protein showed significant activity on Met-Ala-Ser, Met-Gly-Met, Met-Gly-Met-Met and Met-Pro-Gly, but showed no detectable activity on Met-Leu-Phe, Met-His-Arg and Leu-Gly-Gly. These findings led us to conclude that this suppressor gene encodes the second methionine aminopeptidase, and therefore, it was designated as *MAP2*.[60]

Furthermore, a search of GenBank revealed that the suppressor gene is located on chromosome II,[61] and the gene product shares 55% identity in amino acid sequence with that of rat p67. Rat p67 has been shown to be different from the α, β and γ subunits of eIF-2. It can promote protein synthesis in the presence of active eIF-2 kinases by protecting the α-subunit of eIF-2 from eIF-2 kinase-catalyzed phosphorylation, possibly via its *O*-linked GlcNAc residues.[62-66] Recently, we also cloned the gene encoding the human homolog of rat p67. Amino acid sequence comparison revealed that the human homolog shares 92% identity with that of rat p[67]. In addition, the yeast protein, like the rat and human p67, contains highly charged amino acids at the N-terminal region. However, yeast MetAP2 contains only one basic polylysine block, whereas both rat and human p67 contain two polylysine and one aspartic acid rich regions. Other proteins like human eIF-2β and its yeast homolog, Sui3, also contain similar polylysine blocks, which have been postulated to be involved in protein/protein or protein/nucleic acid interactions.[68] It would be of great interest to determine whether these highly charged regions are involved in complex formation and how the differences in their arrangements among these proteins affect their biological functions.

Based on sequence and structure comparison studies, J.F. Bazan, et al have predicted that rat p67 may be a metalloprotease which can modify or inactivate the attacking eIF-2 kinases.[69-70] Their hypothesis differs from the observations of Gupta and coworkers who showed that the eIF-2 kinase was still active in the presence of the rat eIF-2 associated protein, p67, under their assay conditions. This discrepancy remains to be clarified. However, our findings that the yeast homolog of rat p67 is in fact a methionine aminopeptidase led us to propose that the rat p67 may be a mammalian methionine aminopeptidase, and it is possible that yeast MetAP2, like the rat p67, is associated with eIF-2 and has dual functions. The association of MetAP2 with eIF-2 may also be a mechanism for MetAP2 to be associated transiently with the ribosomes to facilitate its N-terminal processing function, and if so, MetAP2 is not only a protein that promotes protein initiation in the presence of active eIF-2 kinases, but also is one of the two key enzymes that are responsible for the removal of the initiator methionine.

To determine the physiological function of the second MetAP in vivo, null mutants were constructed by single step gene disruption. The *map2* null strain has a slow growth rate. These results suggest that disruption of the *MAP2* gene is not lethal. Since the second MetAP may have dual functions, the slow-growth phenotype may be caused by the loss of either its MetAP activity or its function in preventing the phosphorylation of the eIF-2α, or by the loss of both functions. Furthermore, since *map2* null cells grew faster than the *map1* null cells, MetAP1 must be more important than MetAP2 in promoting cell growth.[60] Deletion of both *MAP* genes, followed by tetrad analysis revealed that these two MetAPs are essential for vegetative growth of haploid yeast cells, which further demonstrates that N-terminal processing must provide a critical function for some proteins subjected to this modification.[71-72] These findings also suggest that those nonspecific aminopeptidases in the cytosol, such as AAP1, aminopeptidase yscII, and BLH1, cannot substitute MetAP activity in vivo. Recently, using partial amino acid sequence data derived from porcine MetAP, Arfin et al cloned a full-length cDNA encoding the homologous human enzyme.[73] This cDNA is identical to that encoding the human homolog of rat p67.[67]

We recently successfully overexpressed this human cDNA in insect cells and purified the encoding protein to homogeneity. Activity assays of the purified human protein revealed that the human homolog of rat p67 is indeed a methionine aminopeptidase (Li and Chang, unpublished). These findings strongly suggest that rat p67 is also a mammalian MetAP. It is noteworthy that all mammalian MetAPs discovered to date are homologous to yeast MetAP2. It remains to be established whether there exists a yeast MetAP1 homolog in higher eukaryotes.

CONCLUDING REMARKS

Molecular genetic and biochemical studies of yeast aminopeptidases have provided a great deal of new and interesting information about the biological functions of different aminopeptidases (Table 5.1), such as the possible function of the LAP3 (bleomycin hydrolase) in protecting yeast cells from bleomycin-induced toxic-

Table 5.1. Yeast aminopeptidases[a]

Enzyme	Gene	Location	Knockout phenotype	Bacterial & mammalian homolog
AP yscIII	LAP3 (BLH1)	Cytosol	Bleomycin sensitive	Bleomycin hydrolase
AP yscII	APE2 (LAP1)	Cytosol & Periplasm	No	Aminopeptidase N
Aap1	AAP1	Cytosol	Accumulation of glycogen	Aminopeptidase N
API	LAP4	Vacuole	No	No
APY	APY	Vacuole	No	No
MetAPO1	MAP1	Cytosol	Slow-growth	MetAP
MetAP2	MAP2	cCyosol	Slow-growth	MetAP, p67

[a] Only the yeast aminopeptidases whose encoding genes are cloned are listed.

ity, the regulatory role of Aap1 in the regulation of glycogen accumulation, the cloning of the second methionine aminopeptidase gene, and the finding of its possible role in translational regulation. It is certain that the use of the powerful molecular genetic approaches will allow investigators to study new aminopeptidase genes, to discover novel physiological roles, and to obtain new insights into the structure/function relationships among different aminopeptidases.

ACKNOWLEDGMENTS

The author's work described in this chapter was supported by grants from the Edward Mallinckrodt Jr. Foundation and the National Science Foundation (MCB9512655).

REFERENCES

1. Jones EW. Three proteolytic systems in the yeast *Saccharomyces cerevisiae*. J Biol Chem 1991; 266:7963-66.
2. Rendueles P, Wolf D. Proteinase function in yeast: biochemical and genetic approaches to a central mechanism of post-translational control in the eukaryotic cells. FEMS Microbiol Rev 1988; 54:17-46.
3. Jones EW. The synthesis and function of proteases in *Saccharomyces*: genetic approaches. Annu Rev Genet 1984; 18:233-70.
4. Wilk S, Orlowski MJ. Cation-sensitive neutral endopeptidase: isolation and specificity of the bovine pituitary enzyme. Neurochem 1980; 35:1172-82.
5. Orlowski M. The multicatalytic proteinase complex, a major extralysosomal proteolytic system. Biochemistry 1990; 29:10289-97.
6. Achstetter T, Ehmann C, Osaki A et al. Proteolysis in eukaryotic cells. Proteinase E, a new yeast peptidase. J Biol Chem 1984; 259:13344-48.
7. Kleinschmidt JA, Escher C, Wolf DH. Proteinase yscE of yeast shows homology with the 20 S cylinder particles of *Xenopus laevis*. FEBS Lett 1988; 239:35-40.
8. Enenkel C, Wolf DH. *BLH1* codes for a yeast thiol aminopeptidase, the equivalent of mammalian bleomycin hydrolase. J Biol Chem 1993; 268:7036-43.
9. Caprioglio DR, Padilla C, Werner-Washburne M. Isolation and characterization of AAP1. 1993; 268:14310-15.
10. Carcia-Alvarez N, Cueva R, Suarez-Rendueles P. Molecular cloning of soluble aminopeptidase from *Saccharomyces cerevisiae*. Eur J Biochem 1991; 202:993-02.
11. Nishimura C, Suzuki H, Tanaka N et al. Bleomycin hydrolase is a unique thiol aminopeptidase. Biochem Biophys Res Commun 1989; 163:788-96.

12. Hilt W, Enenkel C, Gruhler A et al. The *PRE4* gene codes for a subunit of the yeast proteasome necessary for peptidylglutamyl-peptide-hydrolyzing activity. Mutations link the proteasome to stress- and ubiquitin-dependent proteolysis. J Biol Chem 1993; 268:3479-86.

13. Fuller RS, Brake A, Thorner J. Yeast prohormone processing enzyme (*KEX2* gene product) is a Ca^{++}-dependent serine protease. Proc Natl Acad Sci USA 1989; 86:1434-38.

14. Sebti SM, Mignano JE, Jani JP et al. Bleomycin hydrolase: molecular cloning, sequencing, and biochemical studies reveal membership in the cysteine proteinase family. Biochemistry 1989; 28:6544-48.

15. Trumbly RJ, Bradley G. Isolation and characterization of aminopeptidase mutants of *Saccharomyces cerevisiae*. J Bacteriol 1983; 156:36-48.

16. Frey J, Rohm KH. Subcellular localization and levels of aminopeptidases and dipeptidases in *Saccharomyces cerevisiae*. Biochim Biophys Acta 1978; 527:31-41.

17. Toh H, Hayashida H, Miyata T. Sequence homology between retroviral reverse transcriptase and putative polymerase of hepatitis B virus and califlower mosaic virus. Nature 1983; 305:827-29.

18. Hirsch HH, Suarez-Rendueles P, Achstetter T et al. Aminopeptidase yscII of yeast. Isolation of mutants and their biochemical and genetic analysis. Eur J Biochem 1988; 173:589-98.

19. Wu Q, Lahti JM, Air GM et al. Molecular cloning of the murine BP-1/6C3 antigen: a member of the zinc-dependent metallopeptidase family. Proc Natl Acad Sci USA 1990; 87:993-97.

20. Watt VM, Yip CC. Amino acid sequence deduced from a rat kidney cDNA suggest it encodes the Zn-dependent aminopeptidase N. J Biol Chem 1989; 264:5480-87.

21. Olsen J, Cowell GM, Konigshofer E et al. Complete amino acid sequence of human intestinal aminopeptidase as deduced from cloned cDNA. FEBS Lett 1988; 238:307-14.

22. Funk CD, Radmark O, Fu JY. Molecular cloning and amino acid sequence of leukotriene A$_4$ hydrolase. Proc Natl Acad Sci USA 1987; 84:6677-81.

23. Bally M, Murgier M, Lazdunski A. Cloning and orientation of the gene encoding aminopeptidase N in *E. coli*. Mol Gen Genet 1984; 195:507-10.

24. Johnson MJ. Isolation and properties of a pure yeast polypeptidase. J Biol Chem 1941; 137:575-86.

25. Chang YH, Smith JA. Molecular cloning and sequencing of genomic DNA encoding aminopeptidase I from *Saccharomyces cerevisiae*. J Biol Chem 1989; 264:6979-83.

26. Cueva R, Garcia-Alvarez N, Suarez-Rendueles P. Yeast vacuolar aminopeptidase yscI. Isolation and regulation of the *APE1* (*LAP4*) structural gene. FEBS Lett 1989; 259:125-29.

27. Metz G, Rohm KH. Yeast aminopeptidase I. Chemical composition and catalytic properties. Biochim Biophys Acta 1976; 429:933-49.

28. Klionsky DJ, Cueva R, Debbie SY. Aminopeptidase I of *Saccharo-*

myces cerevisiae is localized to the vacuole independent of the secretory pathway. J Cell Biol 1992; 119:287-99.

29. Hasilik A, Tanner W. Biosynthesis of the vacuolar yeast glycoprotein carboxypeptidase Y. Conversion of precursor into the enzyme. Eur J Biochem 1978; 85:599-08.

30. Klionsky DJ, Banta LM, Emr SD. Inteacellular sorting and processing of a yeast vacuolar hydrolase: proteinase A propeptide contains vacuolar targeting information. Mol Cell Biol 1988; 8:2105-16.

31. Klionsky DJ, Emr SD. Membrane protein sorting:biosynthesis, transport and processing of yeast vacuolar alkaline phosphatase. EMBO J 1989; 8:2241-50.

32. Banta LM, Robison JS, Klionsky DJ et al. Organelle assembly in yeast: characterization of yeast mutant defective in vacuolar biogenesis and protein sorting. J Cell Biol 1988; 107:1369-83.

33. Chiang HL, Schekman R. regulated import and degradation of a cytosolic protein in the yeast vacuole. Nature 1991; 350:313-18.

34. Chirico WJ, Waters MG, Blobel G. 70K heat shock related proteins stimulate protein translocation into microsomes. Nature 1988; 332:805-10.

35. Deshaies RJ, Scheman R. A yeast mutant defective at an early stage in import of secretory protein precursors into the endoplasmic reticulum. J Cell Biol 1987; 105:633-45.

36. Yauhara T, Nakai T, Ohashi A. Aminopeptidase Y, a new aminopeptidase from *Saccharomyces cerevisiae*. J Biol Chem 1994; 269:13644-50.

37. Nishizawa M, Toshimasa Y, Nakai T et al. Molecular cloning of the aminopeptidase Y gene of *Saccharomyces cerevisiae*. J Biol Chem 1994; 269:13651-55.

38. Duronio RJ, Towler DA, Heuckeroth RO et al. Disruption of the yeast N-myristoyl transferase gene causes recessive lethality. Science 1989; 243:796-800.

39. Gordon JI, Duronio RJ, Rudnick DA et al. Protein N-myristoylation. J Biol Chem 1991; 266:8647-50.

40. Lee FJ, Lin LW, Smith JA. N^α-acetylation is required for normal growth and mating of *Saccharomyces cerevisiae*. J Bacteriol 1989; 171:5795-02.

41. Buss JE, Sefton BM. Direct identification of palmitic acid as the lipid attached to p21ras. Mol Cell Biol 1986; 6:116-22.

42. Carr SA, Biemann K, Shoji S et al. n-Tetradecanoyl is the NH_2-terminal blocking group of the catalytic subunit of cyclic AMP-dependent protein kinase from bovine cardiac muscle. Proc Natl Acad Sci USA 1982; 79:6128-31.

43. Cyert MS, Thorner J. Regulatory subunit (*CNB1* gene product) of yeast Ca^{2+}/calmodulin-dependent phosphoprotein phosphatases is required for adaptation to pheromone. Mol Cell Biol 1992; 12:3460-69.

44. Stearns T, Kahn RA, Botstein D et al. ADP ribosylation factor is

an essential protein in *Saccharomyces cerevisiae* and is encoded by two genes. Mol Cell Biol 1990; 10:6690-99.

45. Stearns T, Willingham MC, Botstein D et al. ADP-ribosylation factor is functionally and physically associated with the Golgi complex. Proc Natl Acad Sci USA 1990; 87:1238-42.

46. Schrock III RD, Lloyd RS. Site-directed mutagenesis of the NH$_2$ terminus of T4 endonuclease V. The position of the alpha NH$_2$ moiety affects catalytic activity. J Biol Chem 1993; 268:880-86.

47. Prchal JT, Cashman DP, Kan YW. Hemoglobin Long Island is caused by a single mutation (adenine to cytosine) resulting in a failure to cleave amino-terminal methionine. Proc Natl Acad Sci USA 1986; 83:24-27.

48. Ben-Bassat A, Bauer K, Chang SY et al. Processing of the initiation methionine from proteins: properties of the *Escherichia coli* methionine aminopeptidase and its gene structure. J Bacteriol 1987; 169:751-57.

49. Miller CG, Strauch KL, Kukral AM et al. N-terminal methionine-specific peptidase in *Salmonella typhimurium*. Proc Natl Acad Sci USA 1987; 84:2718-22.

50. Chang YH, Teichert U, Smith JA. Purification and characterization of a methionine aminopeptidase from *Saccharomyces cerevisiae*. J Biol Chem 1990; 265:19892-97.

51. Kendall RL, Bradshaw RA. Isolation and characterization of the methionine aminopeptidase from porcine liver responsible for the cotranslational processing of proteins. J Biol Chem 1992; 267:20667-73.

52. Moerschell RP, Hosokawa Y, Tsunasawa S et al. The specificities of yeast methionine aminopeptidase and acetylation of amino-terminal methionine in vivo. Processing of altered iso-1-cytochromes C created by oligonucleotide transformation. J Biol Chem 1990; 265:19638-43.

53. Tsunasawa S, Steward JW, Sherman F. Amino-terminal processing of mutant forms of yeast iso-2-cytochrome C. J Biol Chem 1985; 260:5382-91.

54. Huang S, Elliott RL, Liu PS et al. Specificity of cotranslational amino-terminal processing of proteins in yeast. Biochemistry 1987; 26:8242-46.

55. Hirel PH, Schmitter JM, Dessen P et al. Extent of N-terminal methionine excision from *E. coli* proteins is governed by the side-chain length of the penultimate amino acid. Proc Natl Acad Sci USA 1989; 86:8247-51.

56. Roderick SL, Matthews BW. Structure of the cobalt-dependent methionine aminopeptidase from *Escherichia coli*: a new type of proteolytic enzyme. Biochemistry 1993; 32:3907-12.

57. Chang YH, Teichert U, Smith JA. Molecular cloning, sequencing, deletion, and overexpression of a methionine aminopeptidase gene from *Saccharomyces cerevisiae*. J Biol Chem 1992; 267:8007-11.

58. Lovering R, Hanson IM, Borden KL et al. Identification and pre-liminary characterization of a protein motif related to the zinc finger. Proc Natl Acad Sci USA 1993; 90:2112-16.

59. Zuo S, Guo Q, Ling C et al. Evidence that two zinc fingers in the methionine aminopeptidase from *Saccharomyces cerevisiae* are important for normal growth. Mol Gen Gent 1995; 246:247-53.

60. Li X, Chang Y-H. Amino-terminal processing in *Saccharomyces cerevisiae* is an essential function that requires two distinct methionine aminopeptidases. Proc Natl Acad Sci USA 1995; 92:12357-61.

61. Feldman H, Aigle M, Aljinovic G et al. Complete DNA sequence of yeast chromosome II. EMBO J 1994; 13:5795-09.

62. Ray MK, Datta B, Chakaraborty A et al. The eukaryotic initiation factor 2-associated 67 kDa polypeptide (p67) plays a critical role in regulation of protein synthesis initiation in animal cells. Proc Natl Acad Sci USA 1992; 89:539-43.

63. Ray MK, Chakaraborty A, Datta B et al. Characteristics of the eukaryotic initiation factor 2 associated 67 kDa polypeptide. Biochemistry 1992; 32:5151-59.

64. Wu S, Gupta S, Chatterjee N et al. Cloning and characterization of cDNA encoding the eukaryotic initiation factor 2-associated 67 kDa protein (p67). J Biol Chem 1993; 268:10796-01.

65. Datta B, Ray MK, Chakrabarti D et al. Glycosylation of eukaryotic peptide chain initiation factor 2 (eIF-2)-associated 67 kDa polypeptide (p67) and its possible role in the eIF-2 kinase catalyzed phosphorylation of the eIF-2 α-subunit. J Biol Chem 1989; 264:20620-24.

66. Datta B, Chakrabarti D, Roy AL et al. Roles of a 67 kDa polypeptide in reversal of protein synthesis inhibition in heme-deficient reticulocyte lysate. Proc Natl Acad Sci USA 1988; 85:3324-28.

67. Li X, Chang YH. Molecular cloning of a human cDNA encoding an initiation factor 2-associated protein (p67). Biochim Biophys Acta 1995; 1260:333-36.

68. Pathak VK, Nielsen PJ, Trachsel H et al. Structure of the beta subunit of translational initiation factor eIF-2. Cell 1988; 54:633-39.

69. Bazan JF, Weaver LH, Roderick SL et al. Sequence and structure comparison suggest that methionine aminopeptidase, prolidase, aminopeptidase P and creatinase share a common fold. Proc Natl Acad Sci USA 1994; 891:2473-77.

70. Donahue TF, Cigan AM, Pabich EK et al. Mutations at a Zn(II) finger motif in the yeast eIF-2 beta gene alter ribosomal start-site selection during the scanning process. Cell 1988; 54:621-32.

71. Riles L, Olson MV. Nonsense mutations in essential genes of *Saccharomyces cerevisiae*. Genetics 1988; 118:601-07.

72. Taylor A. Aminopeptidases: Towards a mechanism of action. Trends in Biochem 1993; 18:167-72.

73. Arfin SM, Kendall RL, Hall L et al. Eukaryotic methionyl aminopeptidases: Two classes of cobalt-dependent enzymes. Proc Natl Acad Sci 1995; 92:7714-18.

CHAPTER 6

ALTERNATE FUNCTIONS FOR AMINOPEPTIDASES: HYDROLYSIS OF LEUKOTRIENE A$_4$

F. A. Fitzpatrick and Lars Orning

LIPID MEDIATOR BIOSYNTHESIS BY AN AMINOPEPTIDASE ENZYME

This chapter aims to describe an unusual process: the biosynthesis of a lipid mediator of inflammation by an aminopeptidase enzyme. The mediator, leukotriene (LT) B$_4$, is a member of an autacoid family of lipids termed eicosanoids. All eicosanoids originate from enzymatic oxygenation of arachidonic acid. The prostanoids and the leukotrienes compose the two most prominent branches of the eicosanoid family.[1-3] Differences in their enzymology, cellular origins and biological traits help to distinguish the prostanoids from the leukotrienes. The prostanoid family originates from bis-dioxygenase (cyclooxygenase) catalyzed transformation of arachidonic acid into prostaglandin (PG) endoperoxide H$_2$. Spontaneous or enzymatic rearrangement of this labile biosynthetic intermediate generates prostanoids, typified by prostaglandins and thromboxanes. These agents have prominent effects on thrombosis and hemostasis, inflammation, vascular caliber, epithelial integrity, fertility, parturition and other physiological or pathological processes. The leukotriene family originates from a dioxygenase (5-lipoxygenase) catalyzed transformation of arachidonic acid into

Aminopeptidases, edited by Allen Taylor. © 1996 R.G. Landes Company.

LTA$_4$ (5(S)-trans-5,6-oxido-7,9-trans-11,14-cis-eicosatetraenoic acid). Spontaneous or enzymatic hydration of this labile biosynthetic intermediate generates individual leukotrienes, typified by sulfidopeptide leukotrienes (LTC$_4$, LTD$_4$, LTE$_4$) and the dihydroxy leukotriene, LTB$_4$. LTB$_4$ attracts interest as a lipid mediator because it stimulates adhesion of circulating neutrophils to vascular endothelium, directs their migration toward sites of inflammation, and catalyzes the release of their granule constituents. LTB$_4$ also attracts interest because of the unusual enzymology associated with the terminal step in its biosynthesis. This enzymology includes an alternate function for an aminopeptidase enzyme with dual catalytic traits.

Many advances in prostanoid research, spanning the period from 1962 to 1979, preceded the discovery of the leukotriene biosynthetic pathway. For better and for worse leukotriene research often relied on precedents from prostanoid research to design and interpret experiments. These precedents helped identify similarities between the prostanoid and leukotriene pathways of arachidonic acid metabolism. Five key similarities are:

1. Arachidonic acid availability is the rate limiting step at the initiation of cellular eicosanoid biosynthesis. In other words, cellular prostanoid or leukotriene formation begins with activation of phospholipase A$_2$ by receptor-dependent agonists or receptor-independent agonists (e.g., Ca^{2+} ionophore A23187).

2. The cyclooxygenase and lipoxygenase pathways are linear enzymatic pathways with a labile biosynthetic intermediate at a pivotal 'branchpoint' in the pathway.

3. The 'branches' of the cyclooxygenase and lipoxygenase pathways involve spontaneous or enzymatic transformation of the pivotal biosynthetic intermediates.

4. Cellular compartmentalization of specialized enzymes which define the 'branches' of the cyclooxygenase and lipoxygenase pathways are compatible with the functions of the eicosanoid mediators they produce. For instance, leukotriene synthetic enzymes are abundant in leukocytes, and the leukotrienes function as inflammatory mediators.

5. Comprehensive metabolism rapidly inactivates all eicosanoids, as expected for autacoid mediators.

LEUKOTRIENES: CONVERGENCE OF LIPID AND PEPTIDE ENZYMOLOGY

Leukotrienes represent a convergence of lipid enzymology with peptide enzymology.[4-5] This convergence begins with leukotriene biosynthesis (Fig. 6.1). As mentioned above the first steps of leukotriene biosynthesis resemble the first steps of prostanoid biosynthesis. In each pathway: (1) dioxygenase enzymes use the same fatty acid substrate, arachidonic acid; (2) dioxygenase enzymes abstract hydrogen from 1,4-*cis*-pentadiene substituents in a regioselective manner; and (3) dioxygenase enzymes insert the cofactor, O_2, and form a biosynthetic intermediate which distinguishes each pathway. Subsequent steps of leukotriene biosynthesis are unique and involve peptide transferase and dipeptidase enzymes. For example, the pathway leading to LTC_4 involves conjugation of the tripeptide glutathione with the labile biosynthetic intermediate LTA_4. Recent molecular cloning data affirm earlier proposals that the glutathione-S-transferase enzyme which catalyzes the formation of LTC_4 does not resemble glutathione-S-transferases involved in xenobiotic detoxification.[6,7] To form other biologically potent sulfidopeptide leukotrienes, cells exploit two peptidase enzymes. A γ glutamyl peptidase cleaves glutamic acid from LTC_4 to convert it to LTD_4, and a dipeptidase cleaves glycine from LTD_4 to convert it into LTE_4.[8] Thus, the initial steps of the leukotriene pathway, which generate LTA_4, can be considered a variation on a theme used for prostanoid biosynthesis. However, certain terminal steps which govern the formation of sulfidopeptide leukotrienes depart from precedents and use transferase and peptidase enzymes to produce lipopeptide conjugates which initiate, aggravate or sustain hypersensitivity and allergic reactions.[2] With this as background we now consider: what type of terminal enzyme participates in the enzymatic formation of LTB_4?

ENZYMOLOGY OF LTB₄ FORMATION

The terminal step in the enzymology of LTB_4 formation is the catalytic hydration which converts LTA_4, an unstable allylic epoxide, into a stable 5(S),12(R)-dihydroxy-6,14-cis-8,10-trans-eicosatetraenoic acid.[9] It is important to stress the profound differences between spontaneous, nonenzymatic hydration of LTA_4 and enzymatic hydration of LTA_4. Nonenzymatic hydration occurs instantaneously;[10] the nonenzymatic hydration products are a

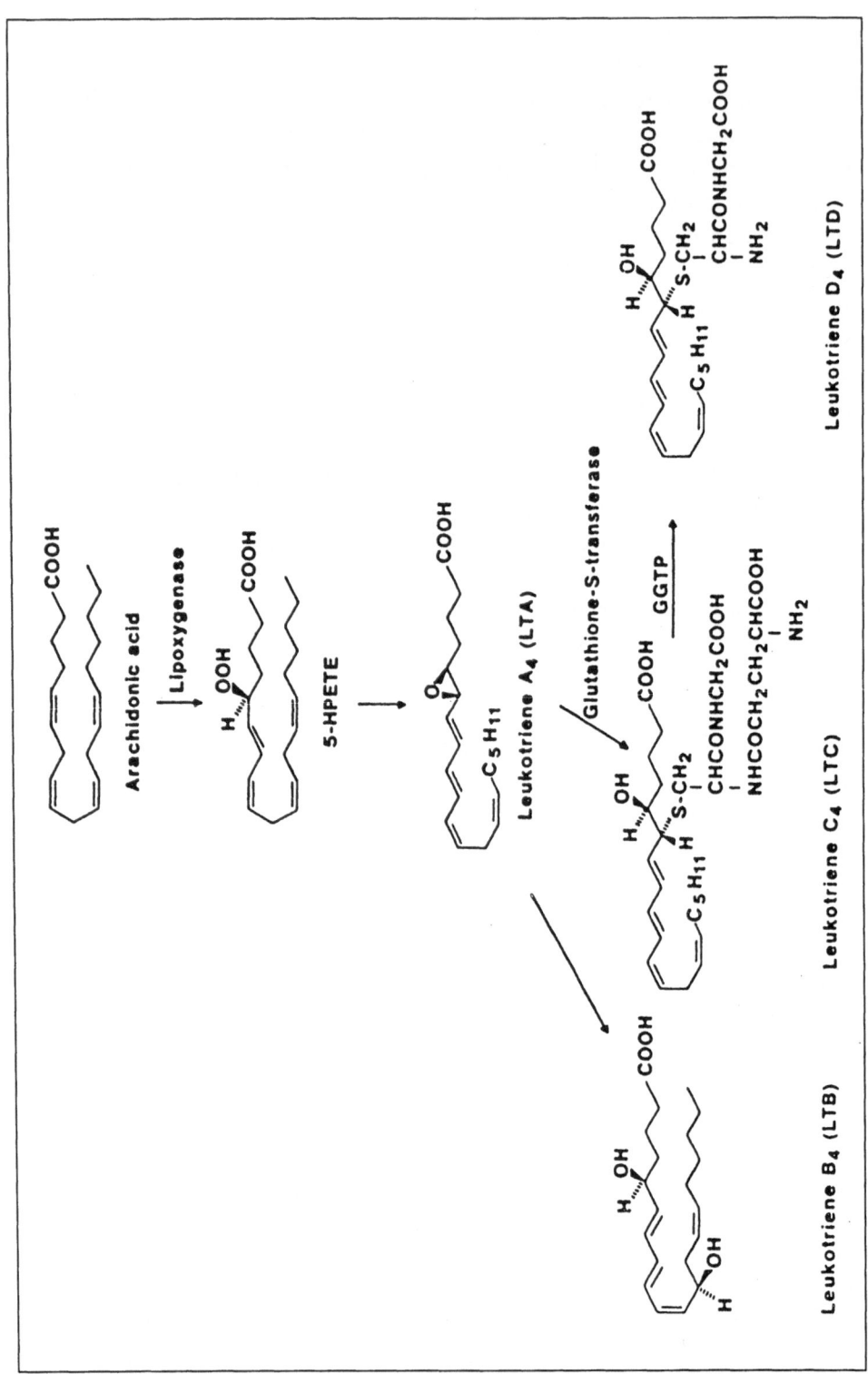

Fig. 6.1. Leukotriene biosynthetic pathway.

pair of 5-(S),12-epi-dihydroxy-eicosatetraenoic acids with olefin bonds at the C6, C8 and C10 positions oriented in an energetically favorable *trans* geometry; the nonenzymatic hydration products exhibit no biological activity. In contrast, enzymatic hydration generates a dihydroxy metabolite with 6-*cis*-olefin geometry and a 5(S)12(R) configuration to produce a potent lipid mediator. Thus, there are profound biological consequences when cells dispose of LTA₄ by enzymatic rather than nonenzymatic hydration. Early investigations on cellular leukotriene biosynthesis fortified this notion by showing that enzymatic hydration was the rate-limiting step in LTB₄ formation and that LTB₄ biosynthesis endured only momentarily, implying a rigorous regulation of cellular LTA₄ hydrolase activity.[11,12]

The availability of synthetic LTA₄ methyl ester allows the in situ generation of LTA₄ for direct investigations of the enzymology and cellular biosynthesis of LTB₄. The availability of the substrate and an enzymatic assay facilitated the isolation, purification and characterization of LTA₄ hydrolase from human and rat leukocytes.[13,14] LTA₄ hydrolase from leukocytes is a soluble, cytosolic protein with a molecular weight of 68,000-70,000 kDa, a broad pH optimum from pH 7-9 and an isoelectric point between pH 5-6. Its molecular weight and biochemical properties suggested that it differed from hepatic epoxide hydrolases.

Two related investigations on the metabolism of intact LTA₄ yielded unexpected and provocative results. First, blood plasma free of any cells converted LTA₄ into LTB₄.[15] Second, a search for the origin of this plasma LTA₄ hydrolase activity revealed that erythrocytes were a novel and unexpected cellular source of LTA₄ hydrolase activity and LTB₄ biosynthesis.[16,17] These results raised two questions. First, if the function of LTA₄ hydrolase was to make LTB₄ why did the enzyme occur in erythrocytes and plasma which had no capacity to generate its substrate LTA₄? Second, if the function of erythrocyte or plasma LTA₄ hydrolase was to make LTB₄ what was the source of its substrate? From one perspective these questions were answered by the discovery that polymorphonuclear leukocytes make and secrete LTA₄ which erythrocytes, platelets and other vascular cells convert efficiently into LTB₄ or LTC₄, respectively.[18-20] This process, termed transcellular biosynthesis, provided a satisfying answer because it corresponded with a precedent from the prostanoid pathway. However, "satisfying" is not necessarily "conclusive". Data available from 1983-1985 might easily have prompted the question: Does the unusual distribution of LTA_4 hydrolase indicate that it has functions distinct from LTB_4 formation? Immunological, immunohistochemical and biochemical reports on its widespread distribution in human and animal tissues did not appear until 1989.[21-23] By 1989 molecular biology was answering the question before it had been posed.

cDNA CLONING OF LTA_4 HYDROLASE

Isolation and cloning of the full-length cDNA encoding LTA_4 hydrolase was the first successful application of molecular biology to an eicosanoid biosynthetic enzyme. Two groups using different approaches and cDNA libraries from different human tissues converged on a solution simultaneously. Funk et al[24] isolated a partial length cDNA for LTA_4 hydrolase by screening a human lung cDNA library with a polyclonal antibody directed against human LTA_4 hydrolase. Availability of the partial length cDNA facilitated the cloning of a full-length cDNA encoding a protein with 610 residues and a predicted molecular weight of 69,140 Da from a placental library. Minami et al[25] isolated a full-length cDNA for LTA_4 hydrolase from a human spleen cDNA library by screening with an oligonucleotide corresponding to a peptide sequence obtained from cyanogen bromide and proteolytic digestion of purified LTA_4 hydrolase. The cDNA from human spleen encoded a protein with 610 residues and a predicted molecular weight of 69,153 Da. Both groups deduced very similar, but not identical, primary amino acid sequences from their nucleotide sequences. Secondary structure predictions and hydropathy plots from both sequences reinforced the conclusion that LTA_4 hydrolase occurred in the cytosol, not the cellular membrane. A hydrophobic region near residues 165-240 drew attention as a potential substrate binding or active site. Most importantly, the nucleotide sequence for LTA_4 hydrolase exhibited no significant homology with sequences for hepatic epoxide hydrolases deposited in the nucleotide data bases at that time, consistent with the conclusions reached by conventional enzymology.[13,14,17] It is informative to consider the impact of this data in 1987. As noted above, the isolation and cloning of LTA_4 hydrolase was the first successful application of molecular biology to an eicosanoid biosynthetic enzyme. In other words, this was the first time that investigators could examine the complete amino acid primary sequence of any eicosanoid biosynthetic enzyme. In retrospect it is not surprising that the complete amino acid sequence of LTA_4 hydrolase prompted no immediate, significant revelations about the unusual nature of this enzyme. In 1987 investigators were encountering for the first time the challenge and opportunity of integrating information derived from classical enzymological approaches with information derived from molecular biological approaches. In 1987 comparison of the sequence of LTA_4 hydrolase with the sequences for epoxide hydrolases was logical: LTA_4 hydrolase catalyzed the hydration of an epoxide substrate. There was no reason to expect that it might be better classified as an aminopeptidase. In addition, data for LTC_4 synthase, another enzyme involved in leukotriene biosynthesis, suggested that the glutathione-S-transferase involved in LTC_4 formation was unique and significantly different from the glutathione-S-transferases involved in xenobiotic conjugation. In 1987 it still seemed reasonable to assume that the terminal enzymes of leukotriene biosynthesis evolved from two well-known enzyme families designed to protect cells from electrophilic attack by either endogenous or exogenous agents. In 1987 the accumulation of conventional biochemical information preceded and guided the use and interpretation of sequence data. This has changed radically: researchers now routinely examine short sequences of cDNA encoding unknown proteins.

RECOGNIZING AN AMINOPEPTIDASE

Recognizing an aminopeptidase catalytic activity within LTA_4 hydrolase occurred in 1989, two years after the publication of its primary sequence in 1987. Recognition depended primarily on the steady accumulation of sequence information for numerous zinc metallohydrolase proteins and on the analysis of the information deposited in protein and nucleotide data bases.[26,27] Jongeneel et al[26] searched the Swiss-Prot data base for the occurrence of the sequence HExxH, a short but critical domain which composes a zinc-binding motif in thermolysin. This sequence occurred in 15 known zinc proteases, as expected. Unexpectedly, this sequence also occurred in over 60 other proteins not necessarily identified as zinc-containing proteins or proteins with peptidase activity. The authors concluded that these proteins might represent previously unrecognized members of a superfamily of zinc containing peptidases. Realignment and analysis of conserved residues flanking the HExxH motif also suggested an extended version of the consensus sequence: VxxHExxH. A few months later Malfroy et al[27] cloned a rat kidney aminopeptidase M and identified it as a zinc containing peptidase which contained the VxxHExxH motif. Searches of data bases revealed several enzymes with substantial homology, plus two enzymes which exhibited weak but detectable homology to aminopeptidase M. These two each contained the VxxHExxH zinc metallohydrolase 'signature' sequence. One of these enzymes was LTA_4 hydrolase:

Zn^{2+} metallohydrolase 'signature' sequence	VxxHExxH
LTA_4 hydrolase	VIAHEISHSWT
Aminopeptidase M_{rat}	VIAHELAHQWF
Aminopeptidase N_{human}	VIAHELAHQWF

Shortly afterwards in a comprehensive review of zinc coordination and the structure-function relationships for zinc enzymes Vallee and Auld[28] drew attention to this observation by stressing the similarity between the zinc binding site of LTA_4 hydrolase, thermolysin and aminopeptidases N and M and by noting that the metal content and peptidase activity of LTA_4 hydrolase were unknown. These reports immediately prompted experiments to determine: (1) Does LTA_4 hydrolase contain zinc ions? (2) Are zinc ions essential for catalysis? (3) Does LTA_4 hydrolase have intrinsic peptidase activity? (4) Do inhibitors of peptidase enzymes also inhibit LTA_4 hydrolase? Investigators were poised to accept a new classification for LTA_4 hydrolase and to discard the notion that it was an unusual epoxide hydrolase because precedents from epoxide hydrolase research had provided few insights into the enzymology of LTA_4 hydrolase. It is interesting to note that eicosanoid researchers were familiar with other examples of a single protein catalyzing two separate and unrelated reactions.[29] Thus, there was ready acceptance of LTA_4 hydrolase as a bifunctional enzyme.

THE ZINC CONTENT AND CATALYTIC TRAITS OF LTA_4 HYDROLASE

By verifying the zinc content of LTA_4 hydrolase, the residues involved in zinc coordination sites, and the role of zinc in the two separate catalytic processes investigators substantiated the hypothesis of Malfroy et al,[27] Vallee and Auld[28] and Toh et al.[30] Using atomic absorption spectrometry Minami et al[31] and Haeggstrom et al[32] established that LTA_4 hydrolase isolated from natural sources contained equimolar amounts of zinc and apoenzyme, consistent with the coordination stoichiometry predicted from primary structure. They also demonstrated that chelating agents, typified by o-phenanthroline, inhibited both catalytic transformation of LTA_4 into LTB_4 and hydrolysis of representative amino acid anilides and naphthylamides. Together with the observation that K_m values for certain amino acid p-nitroanilides were comparable to K_m values reported for aminopeptidase M, it appeared that the chelating agents were inactivating a metalloaminopeptidase and that this aminopeptidase was involved in the conversion of LTA_4 into LTB_4.

Site-directed mutagenesis of key residues within the VxxHExxH motif strengthened the conclusion that LTA_4 hydrolase was a zinc metallohydrolase which catalyzed two separate reactions: (1) hydration of LTA_4 to form LTB_4 a lipid mediator of inflammation and (2) hydrolysis of peptide bonds. Both Minami et al[33] and Wetterholm et al[34] used the crystallographic data from thermolysin to guide their selection of a glutamine residue likely to participate in catalysis by LTA_4 hydrolase/aminopeptidase. Replacement of glu-297 with gln-297 in human LTA_4 hydrolase/aminopeptidase had no effect on LTA_4 catalysis, but it reduced the aminopeptidase activity by >80%. Replacement of glu-297 with ala-, lys- or asp- reduced both catalytic processes. Minami et al[33] verified that the mutations did not modify the protein tertiary structure, and they also identified tyr-234 as a potential proton donor for aminopeptidase catalysis. Wetterholm et al[34] reached similar conclusions for murine LTA_4 hydrolase where glu-296 corresponds to glu-297 of the human enzyme. In addition they reported that mutation of glu-296 did not change the zinc content of the enzyme, consistent with a role for glu-296 in catalysis, not zinc coordination. Subsequent site-directed mutagenesis of his-295, his-299 and glu-318 established unambiguously that these residues coordinate the zinc ion involved in both catalytic processes of the LTA_4 hydrolase/ aminopeptidase.[35] Bacterial expression systems facilitated these investigations with mutated enzymes.[36,37]

PHARMACOLOGICAL INHIBITORS
OF AMINOPEPTIDASE/ LTA$_4$ HYDROLASE

By 1990 the quest for anti-inflammatory agents to inhibit leukotriene formation had led to identification of numerous inhibitors of the 5-lipoxygenase enzyme.[2] In contrast, there were no reversible inhibitors of LTA$_4$ hydrolase, the rate-limiting enzyme for LTB$_4$ biosynthesis. The observation that LTA$_4$ hydrolase shared the unique 'signature' sequence common among Zn^{2+}-metallo-hydrolases prompted us to survey a panel of peptidase inhibitors for their effects on LTA$_4$ hydrolase activity. As mentioned elsewhere in this chapter appreciable data indicated that LTA$_4$ hydrolase differed substantially from known epoxide hydrolase enzymes. First, its lipid substrate requirement was extremely limited. It hydrated only those compounds with a 5,6-oxide-7,9-trans-11,14-cis olefin configuration.[38] It did not hydrate oxiranes which were substrates for hepatic epoxide hydrolases.[17] Second, its product LTB$_4$ was not a vicinal diol, typically formed by epoxide hydrolases, but a dihydroxy compound with a 5(S),12(R) configuration and a 6-cis olefin geometry.[9] Third, known inhibitors of cytosolic hepatic epoxide hydrolase did not inhibit LTA$_4$ hydrolase. Fourth, it exhibited no homology with microsomal or cytosolic epoxide hydrolases deposited in protein or nucleotide data bases.[24,25] Consequently, classification as a peptidase offered a new way to seek inhibitors.

Results showed that bestatin inhibited LTB$_4$ formation by human recombinant LTA$_4$ hydrolase reversibly, with an IC$_{50}$ = 4.0 ± 0.8 μM.[39] Lineweaver-Burk plots indicated a mixture of competitive and noncompetitive mechanisms with an apparent K$_i$ = 201 ± 95 nM. Of the other peptidase inhibitors tested only captopril approached the potency of bestatin (Table 6.1). Bestatin inhibited LTA$_4$ hydrolase selectively without detectable effects on either 5-lipoxygenase or 15-lipoxygenase activity from lysed neutrophils at concentrations 1000-fold higher than its K$_i$. Bestatin also inhibited cellular LTB$_4$ formation by intact erythrocytes and neutrophils.

Consistent with reports by Minami et al[31] and Haeggstrom et al[40] the recombinant human LTA$_4$ hydrolase, purified to homogeneity, contained an intrinsic aminopeptidase activity.[39,41] The rate of hydrolysis of L-leucine-p-nitroanilide was dependent on protein and substrate concentrations with an apparent K_m = 156 μM and a V$_{max}$ = 50 nmol/min/mg enzyme. The reaction rate was constant

Table 6.1. Comparative evaluation of metallohydrolase inhibitors as inhibitors of LTA$_4$ hydrolase

Compound	Concentration (mM) for half-maximal inhibition (IC$_{50}$) activity	
	LTA$_4$ [a]	L-leucine-p-nitroanilide [b]
Bestatin	4 x 10^{-6}	0.3 x 10^{-6}
Amastatin	> 10^{-3}	Inactive[c]
Epibestatin	> 10^{-3}	Inactive
Captopril	11 x 10^{-6}	0.7 x 10^{-7}
Phosphoramidon	Inactive	Inactive
Thiorphan	1 x 10^{-2}	5 x 10^{-6}
Bacitracin	Inactive	Inactive
Glycyl-tyrosine	Inactive	Inactive
L-leucine hydroxamate	Inactive	Inactive
DL-methionine hydroxamate	Inactive	Inactive
o-Phenanthroline	>10^{-1}	Inactive
L-leucine-p-nitroanilide	1 x 10^{-3}	Not determined
L-leucine-β-naphthylamide	3 x 10^{-3}	Not determined
LTA$_4$	Not determined	5 x 10$^{-6(d)}$

[a] LTA$_4$ hydrolase (8 μg/ml) in 0.01 M Tris, pH 8 with BSA (1 mg/ml) and inhibitors were incubated at 25°C for 10 min prior to addition of LTA$_4$ (20 μM). Incubations were continued for 10 min and LTB$_4$ formation was determined by RP-HPLC.
[b] LTA$_4$ hydrolase (14 μg/ml) was incubated with inhibitors and 100 μM L-leucine-p-nitroanilide at 25°C and formation of p-nitroanilide was measured by the increase in absorbance at 405 nm.
[c] No inhibition was detected at the highest concentration examined, 1 mM.
[d] K$_i$

for at least 15 min at 25°C, and, in contrast to the mechanism-based inactivation which accompanied LTB$_4$ formation, there was no mechanism-based inactivation accompanying peptide hydrolysis. A comparison of k_{cat}/K_m values (estimated from V_{max}/K_m) for three substrates (L-leucine p-nitroanilide, L-leucine-β-naphthylamide and LTA$_4$) showed that the LTA$_4$ hydrolase catalyzed LTB$_4$ formation more efficiently than it catalyzed hydrolysis of the two L-leucine substrates. Although aminopeptidase activity and LTA$_4$ hydrolase activity originated from a single protein we found no intrinsic LTA$_4$ hydrolase in aminopeptidase M, suggesting that not all aminopeptidase enzymes were LTA$_4$ hydrolase enzymes. Bestatin also inhibited the LTA$_4$ hydrolase/aminopeptidase catalyzed hydrolysis of L-leucine-p-nitroanilide competitively with K_i = 172 ± 93 nM, a value similar to K_i = 201 ± 95 nM for inhibition of LTB_4 formation. Among other metallohydrolase inhibitors tested, captopril, an inhibitor of angiotensin converting enzyme, was as effective as bestatin.[42]

The identification of bestatin and captopril as inhibitors of LTA_4 hydrolase/aminopeptidase strengthened the conclusion that LTA_4 hydrolase is genetically and functionally related to aminopeptidase M and other Zn^{2+} containing aminopeptidases.[43] Collectively, data accumulated in 1990 and 1991 proved that the family of Zn^{2+} containing peptidases offered better precedents for investigating the inhibition or the catalytic mechanism of LTA_4 hydrolase. For example, precedents from angiotensin converting enzyme led to experiments showing that anions, particularly chloride can activate LTA_4 hydrolase while cations, including Zn^{2+}, can either activate or inhibit LTA_4 hydrolase.[44,45]

Recognition of the catalytic properties of LTA_4 hydrolase/ aminopeptidase and the identification of prototype inhibitors led to more advanced inhibitor designs. A medicinal chemistry research team headed by C-H Wong has contributed exceptionally elegant results.[46] These include the synthesis and characterization of α-keto-β-amino esters and thioamine analogs with sub-micromolar potencies[47-49] and the synthesis and characterization of β-amino-hydroxylamine and aminohydroxamic acids. Certain aminohydroxamic acids are the most potent inhibitors synthesized to date[50] with K_i less than or equal to 1 nM. The hydroxamic acid peptide kelatorphan, and several analogs,[51] also inhibit LTA_4 hydrolase with K_i values near 20 nM. Incorporation of a hydroxamic moiety to chelate active site zinc adds appreciably to the inhibitor potency. For instance, thiorphan, the parent molecule in the kelatorphan series, inhibited peptide hydrolysis with micromolar potency[39] and LTA_4 hydrolysis with millimolar potency. Thiorphan lacks a hydroxamate substituent, and its amide bonds are an important determinant of its activity as an endopeptidase inhibitor.[52] Finally, research groups have devised LTA_4 hydrolase inhibitors by adaptation and incremental modification of prototype LTB_4 antagonists.[53]

A recent investigation by Wong and colleagues[54] offers a detailed view of the binding interactions between LTA_4 hydrolase/ aminopeptidase and 3-(4-benzyloxyphenyl)-2-(R)-amino-propanethiol, an inhibitor with a picomolar K_i value. The data indicate that: (1) thiol on the high affinity inhibitor binds to zinc, displacing water involved in catalytic hydrolysis; and (2) LTA_4 methyl ester inhibits the peptidase activity of the enzyme even though it is a poor substrate for LTA_4 hydrolase. Continued probing of the enzyme with high affinity inhibitors plus the eventual publication of the coordinates of the enzyme crystallized with bestatin will represent a significant advance in our understanding. Two important questions which are unresolved are: (1) does the carboxylic acid of LTA_4 participate in epoxide ring opening? (2) Does zinc at the active site coordinate with the allylic epoxide of LTA_4?

It is uncertain whether selective inhibition of LTA_4 hydrolase in vivo will compare favorably with inhibition of 5-lipoxygenase as a therapeutic tactic. Inhibition of 5-lipoxygenase reduces the formation of both LTB_4 and the sulfidopeptide leukotrienes, all of which have some pro-inflammatory properties. However, selective inhibition of LTA_4 hydrolase may be appropriate in some instances. Until recently the lack of suitable agents thwarted investigation of LTA_4 hydrolase inhibition as a tactic for alleviating inflammation. Another factor accounting for disinterest in selective LTA_4 hydrolase inhibitors as therapeutic agents is a hypothetical concern that they might divert LTA_4 toward increased formation of sulfidopeptide leukotrienes (Fig. 6.1). This could have unpredictable or unfavorable consequences on pulmonary vascular tone or immune cell function. However, it is important to stress that this issue has not been examined experimentally until recently.[55] Identification of bestatin as a prototype LTA_4 hydrolase inhibitor which did not affect 5-lipoxygenase, LTC_4 synthase or phospholipase A_2 provided the opportunity to modulate leukotriene biosynthesis, selectively, and to establish the corresponding effects of this modulation on pulmonary artery perfusion pressure in isolated lungs. Using experimental protocols to compare normal versus inflamed lungs we asked: (1) how much does inhibition of pulmonary LTA_4 hydrolase in vitro divert LTA_4 metabolism away from LTB_4 formation and toward sulfidopeptide leukotriene formation? (2) Does selective modulation of LTA_4 metabolism in vitro have a detrimental effect by increasing pulmonary perfusion pressure? Our results indicate that bestatin inhibits LTA_4 hydrolase selectively, reducing LTB_4 formation with small, and in most cases statistically insignificant effects on sulfidopeptide leukotriene formation. These changes in leukotriene formation had no adverse effects on pulmonary artery perfusion pressure in vitro.[55] As interest in the therapeutic potential of LTA_4 hydrolase inhibitors emerges it will be important to balance their potential value for treatment of inflammatory disorders with potential detrimental effects, especially potential detrimental effects on pulmonary function and detrimental effects originating from nonspecific inhibition of other zinc metallohydrolase enzymes. It is noteworthy that the immuno-modulatory effects of bestatin[56-59] are not directly correlated with inhibition of peptidase.[60] Our data raise the possibility that inhibition of LTA_4 hydrolase also contributes to these effects.

PEPTIDE SUBSTRATES FOR AMINOPEPTIDASE/LTA₄ HYDROLASE

The naturally occurring peptide substrates for aminopeptidase catalysis by LTA_4 hydrolase are still uncertain. We know that LTA_4 is the lipid substrate and that formation of the pro-inflammatory mediator, LTB_4, defines one role for the enzyme. However, the enzyme is also an aminopeptidase whose activation, inhibition, Zn^{2+} content and nucleotide or amino acid sequence resemble other members of a metallohydrolase superfamily. The catalytic versatility of this bifunctional enzyme suggests that it may have a biological role, distinct from LTB_4 formation, involving its proteolytic trait. This implies that hydrolysis of peptide substrates might occur with efficiency and selectivity approaching that of the lipid substrate, LTA_4. Enkephalins are naturally occurring substrates for this enzyme; however, their transformation is inefficient compared to LTA_4.[61] Therefore we systematically investigated the peptide substrate specificity exhibited by recombinant human aminopeptidase/LTA₄ hydrolase (E.C. 3.3.2.6) to clarify its functions.[62] Our data show that tripeptides are 'better' substrates than LTA_4. This is compatible with a biological role for the peptidase activity of the enzyme and may be relevant to the distribution of the enzyme in organs like the ileum, liver, lung and brain.

The binding affinity (K_m) and turnover (k_{cat}) of amino acid p-nitroanilides by LTA_4 hydrolase/aminopeptidase varied according to the amino acid substituent. The specificity constant, k_{cat}/K_m, indicates that recombinant human enzyme preferred L-arginine or L-phenylalanine among the 14 compounds tested. Amino acids with acidic pK_a, with N-terminal substitutions, or with D-stereochemistry were poor substrates (Table 6.2). We surveyed 50 peptides composed of 2 to 11 amino acid residues to clarify the substrate specificity of LTA₄ hydrolase/aminopeptidase. The enzyme displayed highest specificity towards tripeptides, cleaving their amide bond to release the N-terminal amino acid and a corresponding dipeptide. With few exceptions, the enzyme catalyzed hydrolysis of tripeptides with k_{cat}/K_m between 10^5 to 10^6 M^{-1}sec^{-1}. These values equaled or exceeded 1.3×10^5 M^{-1}sec^{-1}, the k_{cat}/K_m for the biologically relevant lipid substrate, LTA₄. The specificity constant declined approximately 10-fold for dipeptides, 100- to 1000-fold for tetra- and pentapeptides and 10,000-fold for hexa- and oligopeptides. The four peptides, RG, RGG, RGD and RGDS, illustrate the relationship between peptide length and substrate specificity. The dipeptide, RG, has a K_m = 2.2 mM and k_{cat} = 23 sec^{-1}. Addition of a third residue, either D- or G-, to the carboxyl terminus enhances substrate binding affinity: K_m = 0.43 mM for RGD and 0.21 mM for RGG. There is a corresponding increase in activity; k_{cat} = 184 sec^{-1} for RGD and 205 sec^{-1} for RGG. Conversely, addition of a fourth residue, S, to the carboxyl terminus reduces substrate binding affinity and activity: K_m = 8.6 mM and k_{cat} = 2 sec^{-1} for RGDS. Data for 11 tripeptides with the formula xGG indicated that the effect of their N-terminal amino acid on catalytic selectivity resembled the effect of the amino acid p-nitroanilides. Catalytic selectivity was maximal for arginine, minimal for valine and glycine and intermediate for leucine, lysine, methionine, alanine and proline.

Data for ten dipeptides with arginine at the N-terminus indicated a preference for an aromatic amino acid at the adjacent position. Most aliphatic amino acids at this position decreased k_{cat}/K_m values 3- to 10-fold; charged amino acids were poorly tolerated. Glycine at the second position yielded a high K_m but also a very high k_{cat}. F appears to be similar to L, K, A, M and P in the tripeptides, but it may be better than these residues in the dipeptide.

Arphamenine A and B are natural analogs of arg-phe and arg-tyr, respectively, and inhibitors of arginine aminopeptidase. Our characterization of the peptide substrate specificity of LTA₄ hydrolase/aminopeptidase, outlined above, allowed us to test the prediction that arphamenines, would also inhibit LTA₄ hydrolase/aminopeptidase. Arphamenine A and arphamenine B inhibited the hydrolysis of several amino acid p-nitroanilides in a reversible, dose-dependent manner. Lineweaver-Burk plots indicated a mixed-

Table 6.2. Kinetic constants for hydrolysis of LTA$_4$ and amino acid p-nitroanilides

	Human recombinant			Mouse recombinant		
	k_{cat}/K_m (sec^{-1}M^{-1})	K_m (mM)	k_{cat} (sec^{-1})	k_{cat}/K_m (sec^{-1}M^{-1})	K_m (mM)	k_{cat} (sec^{-1})
LTA$_4$	127 x 10^3	0.0056	0.71			
Phe-pNA	40 x 10^3	0.23	9.1			
Arg-pNA	37 x 10^3	0.11	4.1	28 x 10^3	0.04	1.10
Ala-pNA	16 x 10^3	0.44	7.2			
Leu-pNA	10 x 10^3	0.26	2.6	10 x 10^3	0.13	1.23
Lys-pNA	9.6 x 10^3	0.048	0.46			
Pro-pNA	9.1 x 10^3	0.22	2.0	11 x 10^3	0.06	0.65
Met-pNA	7.9 x 10^3	0.29	2.3			
Val-pNA	1.3 x 10^3	0.57	0.72			
Gly-pNA	0.36 x 10^3	0.14	0.05			
Asp-pNA	No activity					
Glu-pNA	No activity					
N-benzyl-Arg-pNA	No activity					
N-acetyl-Ala-pNA	No activity					
d-Leu-pNA	No activity					

type inhibition mechanism with lys-p-nitroanilide and pro-p-nitroanilide, but a competitive mechanism with arg-p-nitroanilide. In the latter case, secondary plots of K_m versus inhibitor concentration yielded K$_i$ values of 2.0 μM and 2.5 μM for araphamenine A and B, respectively. Arphamenine A and B inhibited LTB$_4$ formation weakly: IC$_{50}$ values were >300 μM.

Figure 6.2 depicts our model to interpret the interaction of substrates and inhibitors with LTA$_4$ hydrolase/aminopeptidase. S$_1$, S$_1$' and S$_2$' represent three separate subsites on the enzyme which bind residues, designated P$_1$, P$_1$' and P$_2$', of any corresponding ligand. The active site is asymmetric. Since all data indicate that LTA$_4$ hydrolase is also an aminopeptidase it can be assumed to begin at S$_1$. Likewise, the abrupt decrease in k_{cat}/K_m for peptides longer than three residues and the lack of inhibitor effect of tetrapeptide analogs like amastatin suggest that the enzyme lacks an S$_3$' subsite. Interpretation of our data (k_{cat}/K_m) according to this model shows that the enzyme prefers dipeptides and tripeptides with arginine at the N-terminal residue, P$_1$, consistent with the data for p-nitroanilide substrates. Otherwise, substrate specificity is broad, and the S$_1$ subsite can accommodate various amino acids.

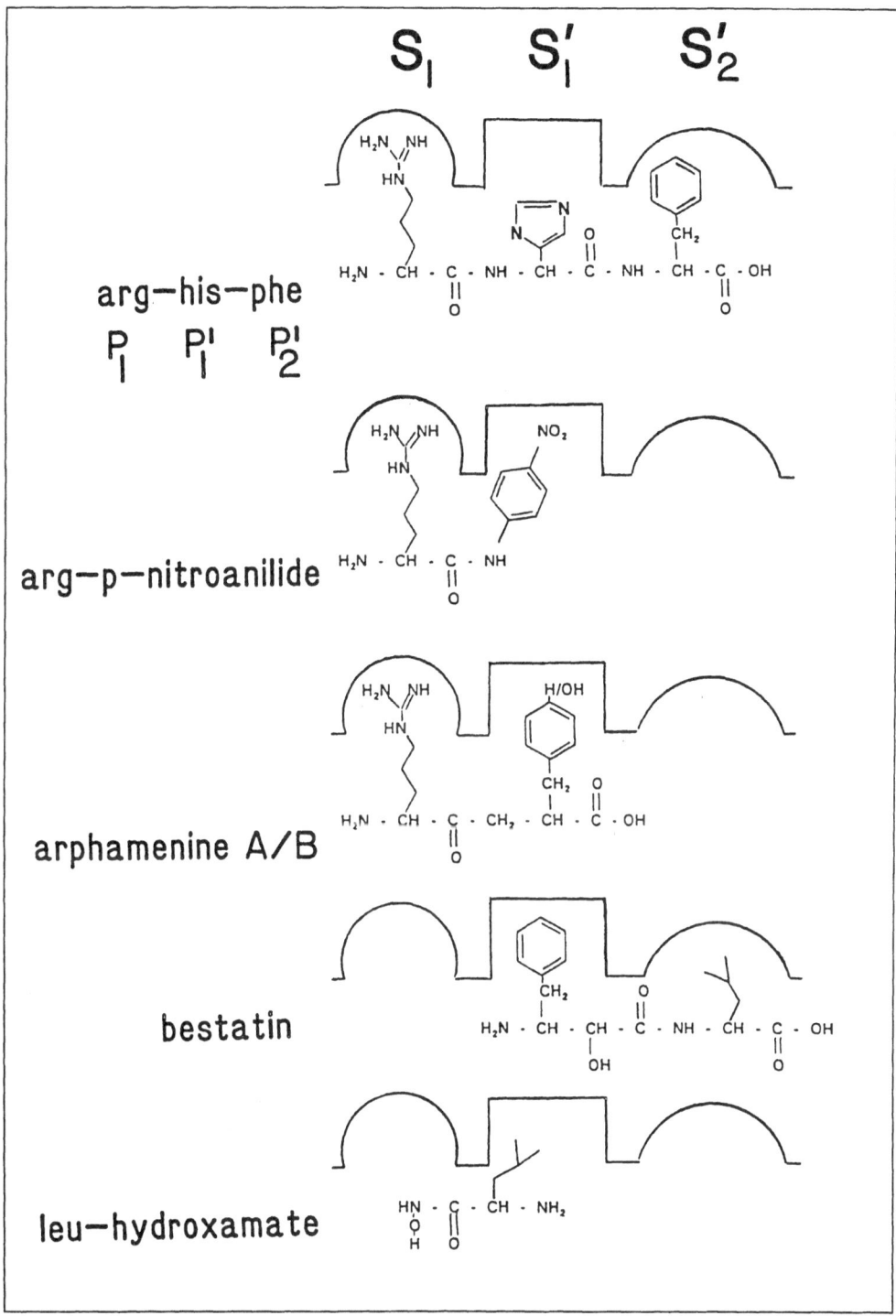

Fig. 6.2. Proposed model for interpreting the interaction of substrates and inhibitors with three subsites (S_1, S_1' and S_2') of aminopeptidase/LTA₄ hydrolase.

Amino acid p-nitroanilides or tripeptides with blocked amino groups at the N-terminus and D-isomers are poor substrates. The low k_{cat}/K_m ratios for asp-p-nitroanilide, glu-p-nitroanilide and glutathione indicate that S_1 does not accommodate acidic residues.

Although L-proline is acceptable at S_1, it is unacceptable at S_1', suggesting a requirement for hydrogen at the amide bond connecting P_1 and P_1'. There is an apparent preference for aromatic or aliphatic amino acids at the P_1' residue. This is evident from the data on dipeptide substrates and from the hydroxamate and arphamenine inhibitors. Arphamenine A and B are analogs of arg-phe and arg-tyr, respectively, and precedents show that they can bind to the S_1 and S_1' subsites of arginine peptidases.[63] Amino acid hydroxamates bind to the S_1' substrate of peptidases.[64] Another potent inhibitor of LTA$_4$ hydrolase/aminopeptidase, bestatin, is a phe-leu analog. Inhibitor binding to arginine aminopeptidase, typified by bestatin, is not necessarily dominant at S_1 and S_1', where peptide and amino acid p-nitroanilide substrates bind. Consistent with a precedent from Harbeson and Rich,[63] investigating a different arginine peptidase, it may involve S_1' and S_2'. Our data for bestatin and 4-nitro-bestatin (K_i = 44 nM, unpublished observation) are also consistent with the apparent preference for aromatic residues at P_1' interacting at the S_1' subsite. Like the S_1 subsite, the S_1' subsite does not accommodate acidic residues. Finally, a small aliphatic residue at P_1' enhances k_{cat} while reducing the binding to S_1'.

Among the peptides tested, including those with acidic, basic and neutral residues there are few restrictions on the P_2' residue. The only requirement seems to be a free carboxyl group. Data suggest a preference for aliphatic residues at P_2'. The high affinity binding of bestatin and nitrobestatin and their proposed interaction at S_1' and S_2' are consistent with this interpretation. Our data are self-consistent with the model proposed in Figure 6.2, and they are also consistent with precedents established by others.[63,64] This model may be useful for design of peptide-based inhibitors of the enzyme. Our model is useful because it can explain the fact that the active sites for peptides and LTA$_4$ substrates overlap, but are not identical. In our model LTA$_4$ would be situated at the S_1' and S_2' subsites because they best accommodate its hydrophobic features and carboxyl group. The positioning of bestatin at the S_1' and S_2' sites is consistent with its potent inhibition of both catalytic processes of the enzyme, while the positioning of arphamenines at S_1 and S_1' is consistent with its potent inhibition of peptide hydrolysis, but not LTA₄ hydrolysis. An interaction of LTA₄ with S_1' and S_2' subsites would be in line with the report that mutating glu-297, positioned at S_1, to gln-297 deletes the aminopeptidase activity but leaves the epoxide hydrolase activity intact.[33,34] Evidence continues to accrue on the amino acid residues which might correspond to the binding sites in our model. For instance, site-directed mutagenesis and chemical modification of tyr-383 eliminated aminopeptidase catalysis but only reduced LTA₄ hydrolase activity by 80-90%, suggesting an area which be an important determinant of the overlap in catalytic processes.[65,66]

CONSTITUTIVE REGULATION OF AMINOPEPTIDASE E.C. 3.3.2.6 BY LTA₄ HYDROLYSIS

Two types of data discussed above support the conclusion that the active site for hydrolysis of LTA₄ overlaps with, but is not identical to, the active site for hydrolysis of peptides. First, site-directed mutagenesis of glu-297 eliminates aminopeptidase activity but not the LTA₄ hydrolase activity of the protein. Second, inhibitors show preferences for either the aminopeptidase or the LTA₄ hydrolase activity, and their preferences are compatible with available models. Data on a constitutive regulatory process which modulates LTA₄ hydrolase activity provide a third source of evidence for a single protein which catalyzes hydrolysis of two types of substrates via a 'shared' active site.

While characterizing LTA₄ hydrolase isolated from erythrocytes we wondered if LTA₄ could react with any nucleophilic residues at or near the active site of the enzyme and inhibit its activity. Kinetic data suggested that such 'suicide' inactivation did accompany catalysis.[17] This process also occurred in intact erythrocytes[67] suggesting that constitutive regulation of LTA₄ hydrolase/aminopeptidase might involve a rare example of 'suicide' inactivation occurring with a naturally occurring substrate-enzyme pair. Using both kinetic and mass spectrometric approaches we demonstrated that 'suicide' inactivation is an efficient, active-site-directed process.[68] Studies with human recombinant LTA₄ hydrolase indicated:

1. A time-dependent loss of LTA₄ hydrolase/aminopeptidase activity by pseudo-first order kinetics.

2. Inactivation was catalysis dependent with a direct relationship between turnover and inactivation.

3. Inactivation was saturable, implying that it proceeded from an enzyme:LTA_4 complex.

4. The pH-dependence for turnover and inactivation were identical indicating a single type of enzyme:LTA_4 complex participating in each pathway.

5. Competitive inhibitors (captopril and bestatin) of LTA_4 hydrolase/aminopeptidase protected against mechanism-based inactivation.

6. Inactivation was irreversible based on the failure to restore enzyme activity after gel filtration and by electrospray mass spectrometry substantiating a covalent addition of LTA_4.

7. There was a 1:1 stoichiometry of inactivation determined by electrospray mass spectrometry. Recently, Mueller et al[70] identified a henicosapeptide encompassing residues 365-385 as the site for binding of LTA_4.

Our kinetic and mass spectrometric data conform with predictions of a conventional model for mechanism-based inactivation (Scheme 1). Basically, mechanism-based inactivation can be viewed as a special case of Michaelis-Menten kinetics in which a single enzyme-substrate complex proceeds along a 'turnover' pathway at a rate designated k_2 in the scheme below, while it proceeds along an inactivation pathway at a rate designated k_3 in the scheme below.

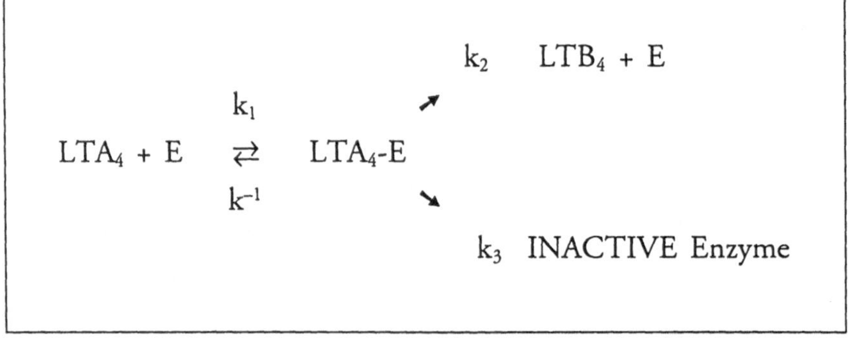

$$LTA_4 + E \quad \underset{k^{-1}}{\overset{k_1}{\rightleftarrows}} \quad LTA_4\text{-}E \quad \overset{k_2}{\nearrow} \quad LTB_4 + E$$
$$\underset{k_3}{\searrow} \quad INACTIVE\ Enzyme$$

Scheme 1. Mechanism-based inactivation of LTA_4 hydrolase/aminopeptidase.

SUMMARY: PAST AND FUTURE

Recognition that an aminopeptidase can use a purely lipid substrate to generate a lipid mediator of inflammation required open minds, the coincidence of molecular biological and biochemical perspectives, and a wealth of material on related peptidases.[43] Recent research suggests that new surprises await us. Successful genomic cloning of LTA₄ hydrolase reveals a 35 kilobase gene with 19 exons localized to chromosome 12q22.[71] Is this a locus for inflammatory diseases with genetic components? We described why it was originally misleading to regard LTA₄ hydrolase/aminopeptidase as a member of the epoxide hydrolase family. Now we encounter xenobiotic response elements in the 5' untranslated region of the LTA₄ hydrolase gene. Is this a constitutive or inducible isoenzyme and what type of transcriptional events follow engagement of this response element? Future experiments will certainly rely on successful strategies employed by others for localizing the binding sites for bestatin and modeling peptide hydrolysis.[72,73]

REFERENCES

1. Smith WL. Prostanoid biosynthesis and mechanisms of action. Am J Physiol 1992; 263:F181-F191.
2. Henderson WR. The role of leukotrienes in inflammation. Ann Intern Med 1994; 121:684-697.
3. Sigal E. The molecular biology of arachidonic acid metabolism. Am J Physiol 1991; 260:L13-L28.
4. Jakschik B, Falkenheim S, Parker C. Precursor role of arachidonic acid in release of slow reacting substances from basophilic leukemia cells. Proc Natl Acad Sci USA 1977; 74:4577-4581.
5. Austen KF. From slow reacting substance of anaphylaxis to leukotriene C₄ synthase. Int Arch Allergy Immunol 1995; 107:19-24.
6. Welsch DJ, Creeley DP, Hauser SD et al. Molecular cloning and expression of human leukotriene C₄ synthase. Proc Natl Acad Sci USA 1994; 91:9745-9749.
7. Lam BK, Penrose JF, Freeman GJ et al. Expression cloning of a cDNA from human leukotriene C4 synthase, an integral membrane protein conjugating reduced glutathione to leukotriene A₄. Proc Natl Acad Sci USA 1994; 91:7663-7667.
8. Anderson ME, Allison RD, Meister A. Interconversion of leukotrienes catalyzed by purified γ glutamyl transpeptidase; concomitant formation of leukotriene D₄ and γ-gutamyl amino acids. Proc Natl Acad Sci USA 1982; 79:1088-1091.
9. Samuelsson B, Funk CD. Enzymes involved in the biosynthesis of

leukotriene B$_4$ J Biol Chem 1989; 264:19469-19472.

10. Fitzpatrick FA, Morton DR, Wynalda MA. Albumin stabilizes leukotriene A$_4$. J Biol Chem 1982; 257:4680-4683.

11. Jakschik B, Kuo CG. Characterization of leukotriene A$_4$ biosynthesis and B$_4$ biosynthesis. Prostaglandins 1983; 25:767-781.

12. Sun FF, McGuire JC. Metabolism of arachidonic acid by human neutrophils. Characterization of the enzymatic reactions that lead to the synthesis of leukotriene B$_4$. Biochim Biophys Acta 1984; 794: 56-64.

13. Radmark O, Shimizu T, Jornvall H et al. Leukotriene A$_4$ hydrolase in human leukocytes: purification and properties. J Biol Chem 1984; 259:12339-12345.

14. Evans JF, Dupuis P, Ford-Hutchinson AW. Purification and characterization of leukotriene A$_4$ hydrolase from rat neutrophils. Biochim Biophys Acta 1985; 840:43-50.

15. Fitzpatrick F, Haeggstrom JZ, Granstrom E et al. Metabolism of leukotriene A$_4$ by an enzyme in blood plasma: a possible leukotactic mechanism. Proc Natl Acad Sci USA 1983; 80:5425-5429.

16. Fitzpatrick FA, Liggett W, McGee JE et al. Metabolism of leukotriene A$_4$ by human erythrocytes. A novel source of leukotriene B$_4$. J Biol Chem 1984; 259:11403-11407.

17. McGee J, Fitzpatrick FA. Enzymatic hydration of leukotriene A$_4$ hydrolase. Purification and characterization of a novel epoxide hydrolase from human erythrocytes. J Biol Chem 1985; 260: 12832-12837.

18. McGee JE, Fitzpatrick FA. Erythrocyte-neutrophil interactions: formation of leukotriene B$_4$ by transcellular biosynthesis. Proc Natl Acad Sci USA 1986; 83:1349-1352.

19. Feinmark SJ, Cannon PJ. Endothelial leukotriene C$_4$ synthesis results from intercellular transfer of leukotriene A$_4$ synthesized by polymorphonuclear leukocytes. J Biol Chem 1986; 261:16466-16472.

20. Maclouf JA, Murphy RC. Transcellular metabolism of neutrophil-derived leukotriene A$_4$ by human platelets. A potential cellular source of leukotriene C$_4$. J Biol Chem 1988; 288:174-181.

21. Bito H, Ohishi N, Miki I et al. Leukotriene A$_4$ hydrolase from guinea pig lung: the presence of two catalytically active forms. J Biochem 1989; 105:261-264.

22. Fu JY, Haeggstrom JZ, Collins P et al. Leukotriene A$_4$ hydrolase: analysis of some human tissues by radioimmunoassay. Biochim Biophys Acta 1989; 1006:121-126.

23. Ohishi N, Minami M, Kobayashi J et al. Immunological quantitation and immunohistochemical localization of leukotriene A$_4$ hydrolase in guinea pig tissues. J Biol Chem 1990; 265:7520-7525.

24. Funk C, Radmark O, Fu JY et al. Molecular cloning and amino acid sequence of leukotriene A4 hydrolase. Proc Natl Acad Sci USA 1987; 84:6677-6681.

25. Minami M, Ohno S, Kawasaki H et al. Molecular cloning of cDNA for human leukotriene A4 hydrolase. J Biol Chem 1987; 262: 13873-13876.
26. Jongeneel CV, Bouvier J, Bairoch A. A unique signature identifies a family of zinc-dependent metallopeptidases. FEBS Lett 1989; 242:211-214.
27. Malfroy B, Kado-Fong H, Gros C et al. Molecular cloning and amino acid sequence of rat kidney aminopeptidase M: a member of a super family of zinc metallohydrolases. Biochem Biophys Res Comm 1989; 161:236-241.
28. Vallee BL, Auld DS. Zinc coordination, function and structure of zinc enzymes and other proteins. Biochemistry 1990; 29:5647-5659.
29. Shimizu T, Radmark O, Samuelsson B. Enzyme with dual lipoxygenase activities catalyzes leukotriene A₄ synthesis from arachidonic acid. Proc Natl Acad Sci USA 1984; 81:689-693.
30. Toh H, Minami M, Shimizu T. Molecular evolution and zinc ion binding motif of leukotriene A4 hydrolase. Biochem Biophys Res Comm 1990; 171:216-221.
31. Minami M, Ohishi N, Mutoh H et al. Leukotriene A4 hydrolase is a zinc-containing aminopeptidase. Biochem Biophys Res Comm 1990; 173:620-266.
32. Haeggstrom JZ, Wetterholm A, Shapiro R et al. Leukotriene A4 hydrolase: a zinc metalloenzyme. Biochem Biophys Res Comm 1990; 172:965-970.
33. Minami M, Bito H, Ohishi N et al. Leukotriene A4 hydrolase, a bifunctional enzyme: distinction of leukotriene A4 hydrolase and aminopeptidase activities by site-directed mutagenesis at glu-297. FEBS Lett 1992; 309:353-357.
34. Wetterholm A, Medina J, Radmark O et al. Leukotriene A4 hydrolase: abrogation of the peptidase activity by mutation of glutamic acid-296. Proc Natl Acad Sci USA 1992; 89:9141-9145.
35. Medina J, Wetterholm A, Radmark O et al. Leukotriene A4 hydrolase: determination of the three zinc binding ligands by site-directed mutagenesis and zinc analysis. Proc Natl Acad Sci USA 1991; 88:7620-7624.
36. Minami M, Minami Y, Emori Y et al. Expression of human leukotriene A₄ hydrolase cDNA in Escherichia coli. FEBS Lett 1988; 229:279-282.
37. Wetterholm A, Medina J, Radmark O et al. Recombinant mouse leukotriene A₄ hydrolase: a zinc metalloenzyme with dual enzyme activities. Biochim Biophys Acta 1991; 1080:96-102.
38. Ohishi N, Izumi T, Minami M et al. Leukotriene A₄ hydrolase in the human lung. Inactivation of the enzyme with leukotriene A₄ isomers. J Biol Chem 1987; 262:10200-10205.
39. Orning L, Krivi G, Fitzpatrick FA. Leukotriene A4 hydrolase inhibition by bestatin and intrinsic aminopeptidase activity establish

its functional resemblance to mextallohydrolase enzymes. J Biol Chem 1991; 266:1375-1378.

40. Haeggstrom JZ, Wetterholm A, Vallee BL et al. Leukotriene A$_4$ hydrolase: an epoxide hydrolase with peptidase activity. Biochem Biophys Res Comm 1990; 173:431-437.

41. Gierse JK, Luckow VA, Askonas LJ et al. High-level expression and purification of human leukotriene A4 hydrolase from insect cells infected with a baculovirus vector. Protein Expression and Purification 1993; 4:358-366.

42. Orning L, Krivi G, Bild G et al. Inhibition of LTA$_4$ hydrolase/aminopeptidase by captopril. J Biol Chem 1991; 266:16507-16511.

43. Taylor A. Aminopeptidases: structure and function. FASEB J 1993; 7:290-298.

44. Orning L, Fitzpatrick FA. Albumin activates peptide hydrolysis by the bifunctional enzyme LTA$_4$ hydrolase/aminopeptidase. Biochemistry 1992;31:4218-4223.

45. Wetterholm A, Haeggstrom JZ. Leukotriene A$_4$ hydrolase: an anion activated peptidase. Biochim Biophys Acta 1992; 1123:275-281.

46. Yuan W, Zhong Z, Wong C-H et al. Probing the inhibition of leukotriene A$_4$ hydrolase based on its aminopeptidase activity. Bioorganic Med Chem Lett 1991; 1:551-556.

47. Yuan W, Wong C-H, Haeggstrom JZ et al. Novel tight-binding inhibitors of leukotriene A$_4$ hydrolase. J Am Chem Soc 1992; 114:6552-6553.

48. Yuan W, Munoz B, Wong C-H et al. Development of selective tight binding inhibitors of leukotriene A$_4$ hydrolase. J Med Chem 1993; 36:211-220.

49. Wetterholm A, Haeggstrom JZ, Samuelsson B et al. Potent and selective inhibitors of leukotriene A$_4$ hydrolase: effects on purified enzyme and human polymorphonuclear leukocytes. J Pharmacol Exptl Therap 1995; 275:31-37.

50. Hogg JH, Ollman IR, Haeggstrom JZ et al. Aminohydroxamic acids as potent inhibitors of leukotriene A$_4$ hydrolase. Bioorganic Med Chem Lett 1995; 3:1405-1415.

51. Penning TD, Askonas LJ, Djuric S et al. Kelatorphan and related analogs: potent and selective inhibitors of leukotriene A$_4$ hydrolase. Bioorganic Med Chem Lett 1995; 5:2517-2522.

52. Monteil T, Kotera M, Duhamel L et al. Importance of the amide bond of thiorphan in the inhibitor-enkephalinase docking process demonstrated with thiorphan isosteres. Bioorganic Med Chem Lett 1992; 2:949-954.

53. Laubadiniere R, Hilboll G, Leon-Lomelli A et al. ω[(ω-Arylalkyl)aryl]alkanoic acid: a new class of specific LTA$_4$ hydrolase inhibitors. J Med Chem 1992; 35:3156-3169.

54. Ollman IR, Hogg H, Munoz B et al. Investigation of the inhibition of leukotriene A$_4$ hydrolase. Bioorganic Med Chem Lett 1995;

3:969-995.

55. Muskardin D, Voelkel N, Fitzpatrick FA. Modulation of pulmonary leukotriene formation and perfusion pressure by bestatin, an inhibitor of leukotriene A4 hydrolase. Biochem Pharmacol 1994; 48:131-137.

56. Ishizuka M, Masuda T, Kabayashi N et al. Effect of bestatin on mouse immune system and experimental murine tumors. J Antibiotics 1980; 33:642-652.

57. Ishizuka M, Sato J, Sugiyama Y et al. Mitogenic effect of bestatin on murine lymphocytes. J Antibiotics 1980; 33:653-662.

58. Umezawa H, Ishizuka M, Aoyagi T et al. Enhancement of delayed type hypersensitivity by bestatin, an inhibitor of aminopeptidase. J Antibiotics 1976; 29:857-859.

59. Umezawa H. Studies on low molecular weight immunomodifiers: bestatin discovery and actions. Drugs Exptl and Clin Res 1984; 10:519-531.

60. Wilkes S, Prescott J. The slow, tight-binding of bestatin and amastatin to aminopeptidases. J Biol Chem 1985; 260:13154-13162.

61. Griffin K, Gierse J, Krivi G et al. Opioid peptides are substrates for the bifunctional enzyme LTA₄ hydrolase/aminopeptidase. Prostaglandins 1992; 4:251-257.

62. Orning L, Gierse JK, Fitzpatrick FA. The bifunctional enzyme leukotriene A₄ hydrolase is an arginine aminopeptidase of high efficiency and specificity. J Biol Chem 1994; 269:11269-11273.

63. Harbeson SL, Rich DH. Inhibition of arginine aminopeptidase by bestatin and arphamenine analogs. Evidence for a new mode of binding to aminopeptidases. Biochemistry 1988; 27:7301-7310.

64. Tronrud DE, Roderick SL, Mathews BW. Structural basis for the action of thermolysin. Matrix 1992; supplement 1:107-111.

65. Blomster M, Wetterholm A, Mueller MJ et al. Evidence for a catalytic role of tyrosine in the peptidase reaction of leukotriene A₄ hydrolase. Eur J Biochem 1995; 231:528-534.

66. Mueller M, Samuelsson B, Haeggstrom JZ. Chemical modification of leukotriene A₄ hydrolase. indications for essential tyrosinyl residues at the active site. Biochemistry 1995; 34:3536-3543.

67. Orning L, Jones D, Fitzpatrick FA. Mechanism-based inactivation of leukotriene A4 hydrolase during leukotriene B4 formation by erythrocytes. J Biol Chem 1990; 265:14911-14916.

68. Orning L, Gierse J, Duffin K et al. Mechanism-based inactivation of leukotriene A₄ hydrolase/aminopeptidase by leukotriene A₄: mass spectrometric and kinetic characterization. J Biol Chem 1992; 267:22733-22739.

69. Evans JF, Kargman S. Bestatin inhibits covalent coupling of [³H] LTA₄ to human leukocyte LTA₄ hydrolase. FEBS Lett 1992; 297:139-142.

70. Mueller MJ, Wetterholm A, Blomster M et al. Leukotriene A₄ hy-

drolase: mapping of a henicosapeptide involved in mechanism-based inactivation. Proc Natl Acad Sci USA 1995; 92: 8383-8387.

71. Mancini J, Evans JF. Cloning and characterization of the human leukotriene A$_4$ hydrolase gene. Eur J Biochem 1995; 231:65-71.

72. Burley SK, David PR, Lipscomb WN. Leucine aminopeptodase: bestatin inhibition and a model for enzyme-catalyzed peptide hydrolysis. Proc Natl Acad Sci USA 1991; 88:6916-6920.

73. Taylor A, Peltier CZ, Jahngen EGE et al. Use of azidobestatin as a photoaffinity label to identify the active site peptide of leucine aminopeptidase. Biochemistry 1992; 31:4141-4150.

PHYSIOLOGICAL ROLES OF ECTOENZYMES INDICATED BY THE USE OF AMINOPEPTIDASE INHIBITORS

Takaaki Aoyagi

INTRODUCTION

Most biological processes, from ontogenesis through death, in normal or diseased conditions, are catalyzed by specific enzymes. Intrinsic inhibitors of these enzymes also play an important role in controlling some of these processes. In order to elucidate the role of aminopeptidases in diseases and to design aminopeptidase-based therapeutics, it is essential to understand the mechanisms of action of these enzymes and the physiological roles of the inhibitors.

For some physiological processes, aminopeptidases and other enzymes work in concert. We have studied the interaction between these processes, with particular emphasis on elucidating relationships between aminopeptidases activity and disease. To do this, we search the filtrate of microbial culture broth for inhibitors of target enzymes, the goal being to discover the most effective therapeutics against the diseases. We have given particular attention to inflammation, cancer, immune disorders, diabetes, hypertension,

Aminopeptidases, edited by Allen Taylor. © 1996 R.G. Landes Company.

hyperlipemia, viral infections and Alzheimer's disease. As part of these investigations, we discovered a series of low-mass inhibitors of critical enzymes.[1-9]

Ectoenzymes, the enzymes which are located on the cellular surface, have attracted special interest recently (see chapter 1). Included in ectoenzymes are aminopeptidases, alkaline phosphatase, and esterases.[10-12] The enzyme inhibitors, which have been discovered in microbial metabolites, have peculiar characteristics in common: they are of low mass, low toxicity and with a unique chemical structure. These inhibitors are useful in the analysis of reaction mechanisms, including stereospecificity of the corresponding enzymes. They have also been used as affinity chromatography ligands to purify enzymes and proenzymes. Furthermore, some inhibitors of cell-surface enzymes show immunomodulatory activity and are related to the metabolism of various physiologically active peptides.[9]

MEMBRANE AMINOPEPTIDASES

Inhibition of growth or locomotion of mammalian cells, their morphological changes, and metastasis are mediated through their mutual recognition and interactions between cellular membranes and cellular functions.[13] During the early stage of infection with influenza virus, we noticed a rearrangement of sialic acid-containing glycoconjugates of the plasma membrane of host cells. At the same time, the fluidity of plasma membrane is decreased when virion is absorbed on the cell membrane. This causes a drastic change in the activities of the enzymes which are located on the cell membrane.[10,14,15]

Furthermore, we determined the enzymatic activities on myxovirus virion grown in cholioallantoic membrane (CAM) cells and in monkey kidney (MK) cells. In case of such virions grown in CAM cells, the activity of aminopeptidase B (AP-B) is higher than AP-A. Similar relationships describe activities on the surface of the host cells. However, in the virion grown in MK cells, and on the surface of the MK cells, the activity of AP-A is higher than AP-B.[16,17] This suggests that the aminopeptidase in the envelope of the virion came from that located on the plasma membrane of the host cells. These results indicate that the surface of plasma membrane has various enzymes located on it. We have found the activities of aminopeptidases, alkaline phosphatase, and esterase on the surface of lymphocytes, macrophages, and on normal or virus-transformed and tumor cells.

AMINOPEPTIDASES IN MAMMALIAN SERUM AND ORGANS

Aminopeptidases (AP) are defined as a group of enzymes which specifically cleave peptide bonds near amino-terminals of polypeptides. Table 7.1 compiles AP from mammalian sources with their E.C. number[18] (see chapter 1). Many of them are metalloenzymes with metals coordinated in their active centers, but the others use cysteine and serine at the active site.[19,20]

AP-N (AP-M, arylamidase, acyl-peptide hydrolase) is an aminopeptidase which has been the subject of a number of studies. The most favorable substrates are peptides having Ala at their amino-terminals; the enzyme can also rapidly release Leu, Phe, Tyr, Arg, Met, Lys, Trp, Gly, Gln, Ser and His. However, if the amino-terminal is Pro or pyroglutamyl-, rates of catalysis are limited.[21]

AP-N is buried in cellular membrane, with its carboxyl-terminal segment outside the cells and with its amino-segment inside the cells. The enzyme is abundantly distributed on the cellular surface of the distant part in renal tubules, epithelial cells, and microvilli in the small intestine. The enzyme is also expressed on the surface of monocytes and granulocytes and is identical with CD13, a specific marker of these blood cells.[22] The level of soluble AP-N in serum increases remarkably in hepatic and pancreatic tumors. AP-N activity was also reported to increase in hepatic diseases induced by excessive uptake of alcohol or drugs.[19,23] Aminopeptidases are also involved in biosynthesis and degradation of enkephalin[24,25] and thymopentin,[26] thus affecting the pharmacological activities of these peptides.

AMINOPEPTIDASE INHIBITORS

Microbes which produce aminopeptidase inhibitors, as well as chemical structure of the latter, are shown in Table 7.2 and Figure 7.1. Amastatin strongly inhibits AP-A, AP-N, Leu-AP, Tyr-AP and tripeptidyl and tetrapeptidyl aminopeptidases.[27] Amastatin has five asymmetric carbon atoms in its molecules. Bestatin (also known as ubenimex) strongly inhibits AP-B, AP-N, Ala-AP, Leu-AP, and tripeptidyl and tetrapeptidyl aminopeptidases[28] (see chapter 2). Bestatin has three asymmetric centers, and its eight stereoisomers

Table 7.1. Aminopeptidase in mammalian sources

Enzyme	E.C.	Active site	Substrate specificity	MW (kDa)
AP-N*	3.4.11.2	Zn	$\downarrow$ X—Y— (X=many)	160
AP-A*	3.4.11.7	Ca	$\downarrow$ X—Y— (X=Glu,Asp)	170
AP-W*	3.4.11.-	Zn	$\downarrow$ X—Y— (X=Trp, Phe, Tyr)	130
AP-P*	3.4.11.9	Unknown	$\downarrow$ X—Y— (Y=Pro)	100
AP-B	3.4.11.6	Unknown	$\downarrow$ X—Y— (X=Lys,Arg)	30-100
Leu-AP	3.4.11.1	Zn	$\downarrow$ X—Y— (X=many)	53,65
PyrGlu-API**	3.4.19.3	Unknown	$\downarrow$ X—Y— (X=PyrGlu)	72
Cys-AP	3.4.11.3	Unknown	$\downarrow$ X—Y— (X=Cys,Leu)	32
Pro-iminopeptidase	3.4.11.5	Unknown	$\downarrow$ X—Y— (X=Pro)	
Dipeptidylpeptidase IV	3.4.14.5	Serine	$\downarrow$ X—Y—Z— (X=many; Y=Ala)	130
Dipeptidase	3.4.13.-	Zn	$\downarrow$ X—Y (X,Y=many)	50
Tripeptide aminopeptidase	3.4.19.3	Cysteine	$\downarrow$ X—Y̅—Y (X=Gly,Leu; Y-Gly)	130

*: cell surface binding enzyme
**: pyroglutamylaminopeptidase inhibitor

Table 7.2. Strains of microbes producing aminopeptidase inhibitors

Inhibitor	Inhibited enzyme	E.C.	Strain
Amastatin	AP-A	3.4.11.7	*Streptomyces sp.* ME98-M3
Bestatin	AP-B	3.4.11.6	*S. olivoreticuli*
Arphamenines A and B	AP-B	3.4.11.6	*Chromobacterium violaceum*
Aminoacylarginine	AP-B	3.4.11.6	*Actinomyces* MF931-A2
Antibiotic OF 4949-I, II, III, IV	AP-B	3.4.11.6	*Penicillium rugulosum*
Actinonin	AP-N	3.4.11.2	*Streptomyces sp.* MG848-hF6
Probestin	AP-N	3.4.11.2	*S. azureus* MH663-2F6
Prostatin	AP-N	3.4.11.2	*S. azureus* MH663-2F6
Leuhistin	AP-N	3.4.11.2	*B. laterosporus* BMI156-14F1
AHPA-Val-Phe[1]	AP-N	3.4.11.2	*Streptomyces sp.* MJ716-m3
Hydrostatins A and B	Leu-AP	3.4.11.1	*Streptomyces sp.* MK7-NF1
DHBS[2]	Leu-AP	3.4.11.1	*Actinomyces* MJ946-SF6
Tyromycin A	Leu-AP	3.4.11.1	*Tyromyces lacteus*
Pyrizinostatin	PyrGlu-P[3]	3.4.19.3	*Streptomyces sp.* SA2289
2-Me-fervenulone	PyrGlu-P	3.4.19.3	*Streptomyces sp.* SA2289
Benarthin	PyrGlu-P	3.4.19.3	*S. xanthophaeus* MJ244-SF1
Formestins A and B	fMet-AP	3.4	*Actinomadura reseoviolacea*
Ebelactones A and B	fMet-AP	3.4	*S. aburaviensis*

(1) (2S,3R)-3-amino-2-hydroxy-4-phenylbutanoyl-L-valyl-L-phenylalanine
(2) N-(2,3-Dihydroxybenzoyl)-serine, bimol, ester
(3) Pyroglutamyl peptidase

have been synthesized. The S-configuration of the C2 of the (2S,3R)-3-amino-2-hydroxy-4-phenylbutanoyl moiety (AHPA) is required for activity. All the four stereoisomers of bestatin with 2S-configuration display inhibitory activities against AP-B and Leu-AP, whereas the four other stereoisomers with 2R-configuration have no, or only slight, activities.[29,30] The immune-enhancing effect of bestatin has been the subject of considerable study.

Arphamenines are specific inhibitors of AP-B.[31] The structures of these inhibitors have a methylene ketone (-CO-CH$_2$-) in place of the scissile peptide bond (-CO-NH-) of L-arginyl-L-phenylalanine, the substrate of AP-B (Fig. 7.1).[32] Many reports have been published regarding the biosynthesis of peptide bonds of various peptide compounds. None reported the biosynthesis of a methylene ketone group linking two amino acids. We determined the biosynthetic pathway of arphamenine A.[33] Furthermore, we also found L-isoleucyl-L-arginine, L-leucyl-L-arginine, and L-valyl-L-arginine by screening for AP-B.[34]

Recently, it was reported that glycoprotein CD13 (gp150), one of the differentiation antigens on the surface of leukocytes, is identical with AP-N (AP-M)[35] (see chapter 2). We have discovered many inhibitors of AP-N, such as probestin,[36,37] leuhistin,[38,39] actinonin,[40] amastatin[27] and bestatin.[28] Bestatin binds to the surface of macrophages at the ratio of 10^7 molecules per cell.[3]

N-Formylmethionine aminopeptidase (fMetAP) is also located on the cell surface and has been reported to play an important role in chemotaxis. We were successful in obtaining inhibitors of fMetAP, which we named formestins A and B (Table 7.2, Fig. 7.1).[9] Ebelactones A and B,[41,42] which were found by screening for inhibitors of esterase and lipase, inhibited not only esterase

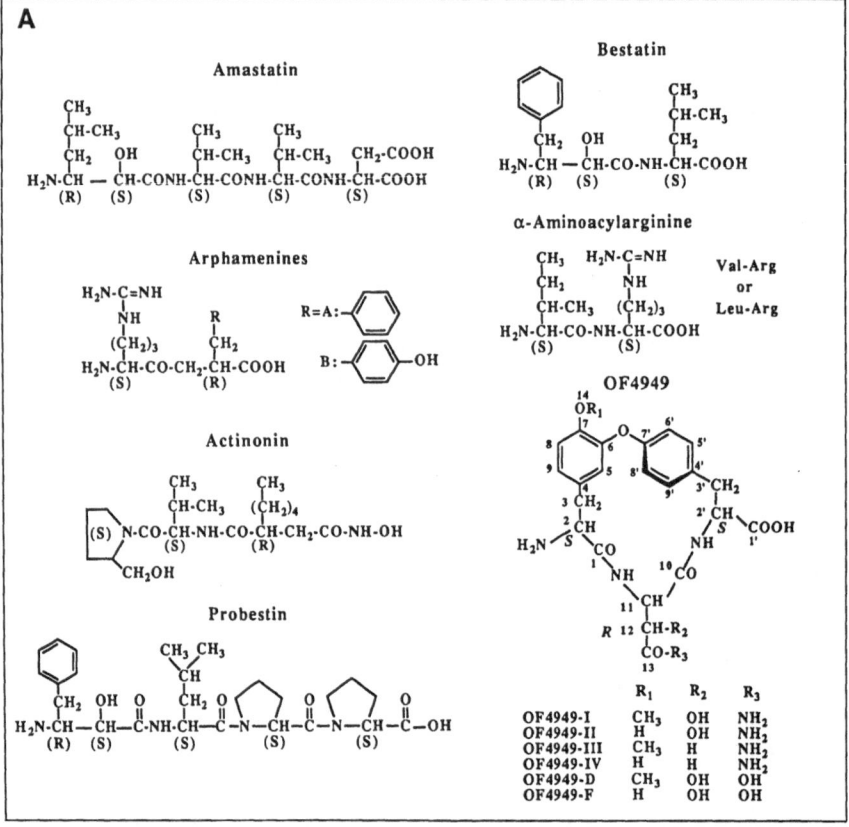

Fig. 7.1. Structure of aminopeptidase inhibitors. In (B) DHBS = N-(2,3-Dihydroxybenzoyl)-serine, bimol, ester; AHPA-Val-Phe = (2S,3R)-3-amino-2-hydroxy-4-phenylbutanoyl-L-valyl-L-phenylalanine.

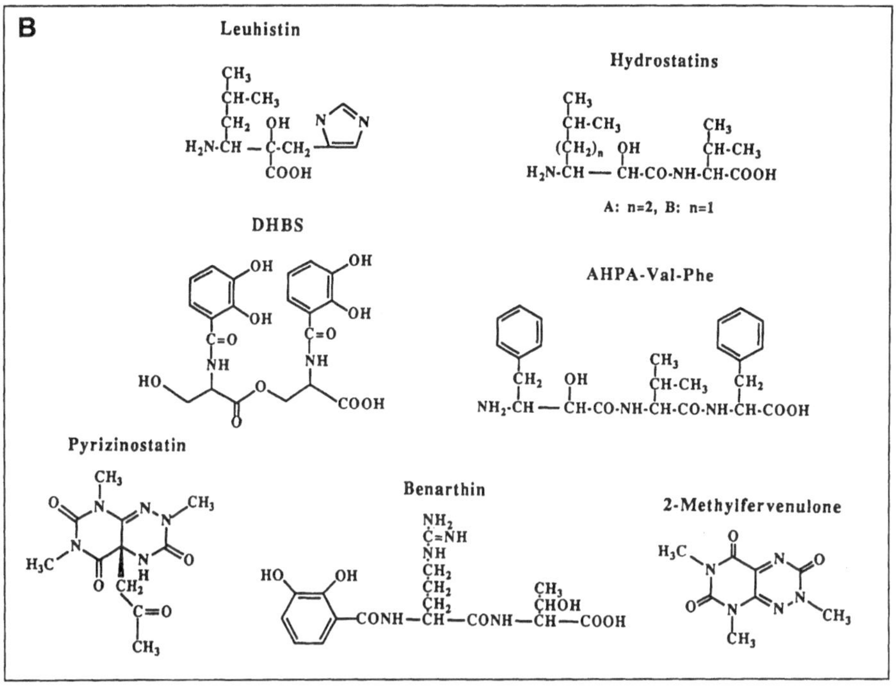

B

Leuhistin

Hydrostatins

$$\text{H}_2\text{N-CH} - \underset{\underset{\text{COOH}}{|}}{\overset{\overset{\text{CH}_3}{\overset{|}{\text{CH-CH}_3}}}{\overset{|}{\underset{|}{\text{CH}_2}}}}\overset{\text{OH}}{\underset{|}{\text{C}}}\text{-CH}_2$$

$$\text{H}_2\text{N-CH} \overset{\overset{\text{CH}_3}{\overset{|}{\text{CH-CH}_3}}}{\underset{\underset{\text{CH-CO-NH-CH-COOH}}{|}}{\overset{|}{\underset{|}{(\text{CH}_2)_n}}}}\quad\overset{\text{CH}_3}{\overset{|}{\text{CH-CH}_3}}$$

A: n=2, B: n=1

DHBS

AHPA-Val-Phe

Pyrizinostatin

Benarthin

2-Methylfervenulone

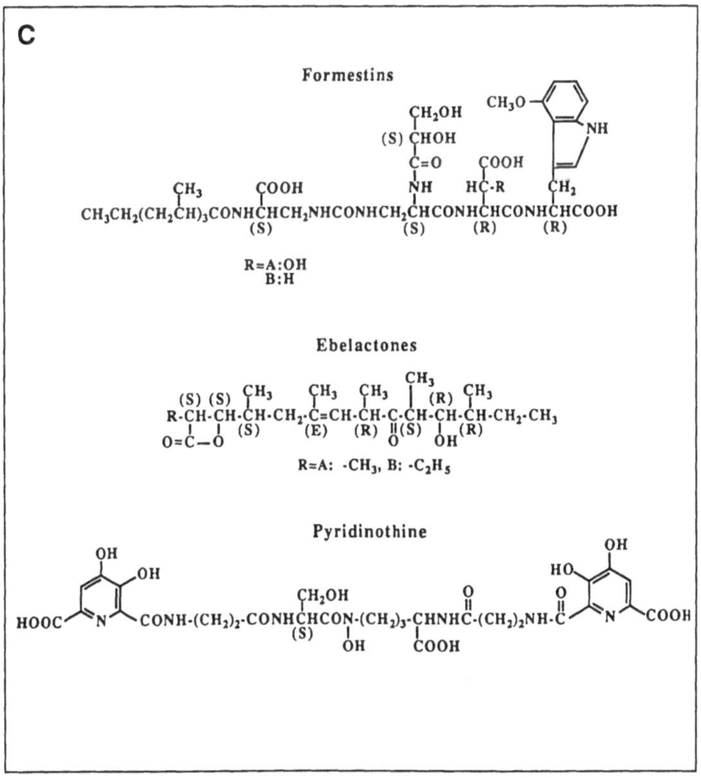

C

Formestins

$$\text{CH}_3\text{CH}_2(\text{CH}_2\text{CH})_3\text{CONHCHCH}_2\text{NHCONHCH}_2\text{CHCONHCHCONHCHCOOH}$$

R=A:OH
B:H

Ebelactones

$$\text{R-CH-CH-CH}_2\text{-C=CH-CH-C-CH-CH-CH-CH}_2\text{-CH}_3$$

R=A: -CH$_3$, B: -C$_2$H$_5$

Pyridinothine

$$\text{HOOC}\quad\text{CONH-(CH}_2)_2\text{-CONHCHCON-(CH}_2)_3\text{-CHNHC-(CH}_2)_2\text{NH-C}\quad\text{COOH}$$

and lipase but also fMetAP. In both esterastin[43] and ebelactone A and B, the β-lactone part of the molecules is important for inhibitory activity (Table 7.3, Fig. 7.1). The K_i values and the type of inhibition of these inhibitors against exopeptidases are shown in Table 7.3 (see chapter 2).

An inhibitor of dipeptidylaminopeptidase (DPP-I) was identified as 2,3-dicarboxyaziridine (DCA). Specific inhibitors of DPP-II, epostatin (unpublished data), and dioctatins A and B, were also discovered.[9] We also discovered fluostatins A and B, inhibitors of DPP-III (unpublished data). Recently, it was reported that glycoprotein CD26, a differentiation antigen on leukocytes' surfaces, is identical with DPP-IV.[44-46] We screened for an inhibitor against DPP-IV and discovered diprotins A and B.[47] The microbes producing these inhibitors and their chemical structures are shown in Table 7.4 and Figure 7.2. The K_i values and the type of inhibition of these inhibitors against DPP are shown in Table 7.5.

Table 7.3. Kinetic constants of aminopeptidase inhibitors

Inhibitor	Enzyme	Substrate	*Km* ($\times 10^{-4}$M)	K_i ($\times 10^{-7}$M)	Type of inhibition
Amastatin	AP-A	Glu·NA[(1)]	8.0	2.5	Competitive
Amastatin	Leu-AP	Leu·NA	37.0	16.0	Competitive
Bestatin	AP-B	Arg·NA	1.0	0.6	Competitive
Bestatin	Leu-AP	Leu·NA	5.8	0.2	Competitive
Arphamenine A	AP-B	Arg·NA	1.0	0.025	Competitive
Arphamenine B	AP-B	Arg·NA	1.0	0.008	Competitive
Ile-Arg	AP-B	Arg·NA	1.0	5.0	Competitive
Val-Arg	AP-B	Arg·NA	1.0	21.0	Competitive
Actinonin	AP-N(M)	Leu·NA	0.8	1.7	Competitive
Probestin	AP-N(M)	Leu·NA	0.8	0.19	Competitive
Leuhistin	AP-N(M)	Leu·NA	0.8	2.3	Competitive
Hydrostatin A	Leu-AP	Leu·NA	13.0	4.0	Competitive
Hydrostatin B	Leu-AP	Leu·NA	13.0	20.0	Competitive
AHPA-Val-Phe[(2)]	AP-N	Leu·NA	0.8	0.295	Competitive
Pyrizinostatin	PyrGlu-P[(3)]	PyrGlu·NA	0.33	—	Noncompetitive
Benarthin	PyrGlu-P	PyrGlu·NA	0.33	12.0	Competitive
Formestin A	fMet-AP	fMet·NA	2.0	0.39	Competitive
Formestin B	fMet-AP	fMet·NA	2.0	1.23	Competitive
Ebelactone A	fMet-AP	fMet·NA	2.0	1.73	Noncompetitive
Ebelactone B	fMet-AP	fMet·NA	2.0	0.63	Noncompetitive

(1) L-glutamic acid β-naphthylamide
(2) (2S,3R)-3-amino-2-hydroxy-4-phenylbutanoyl-L-valyl-L-phenylalanine
(3) Pyroglutamyl peptidase

Table 7.4. Strains of microbes producing dipeptidylpeptidase inhibitors

Inhibitor	Inhibited enzyme	Strain
DCA[1]	DPP-I	*Streptomyces sp.* MD398-A1
Epostatin	DPP-II	*Streptomyces sp.* MJ995-OF5
Dioctatins A and B	DPP-II	*Streptomyces sp.* SA-2581
Fluostatins A and B	DPP-III	*Streptomyces sp.* TA-3391
Diprotins A and B	DPP-IV	*Bacillus cereus* BMF673-RF1

(1) 2,3-Dicarboxyaziridine

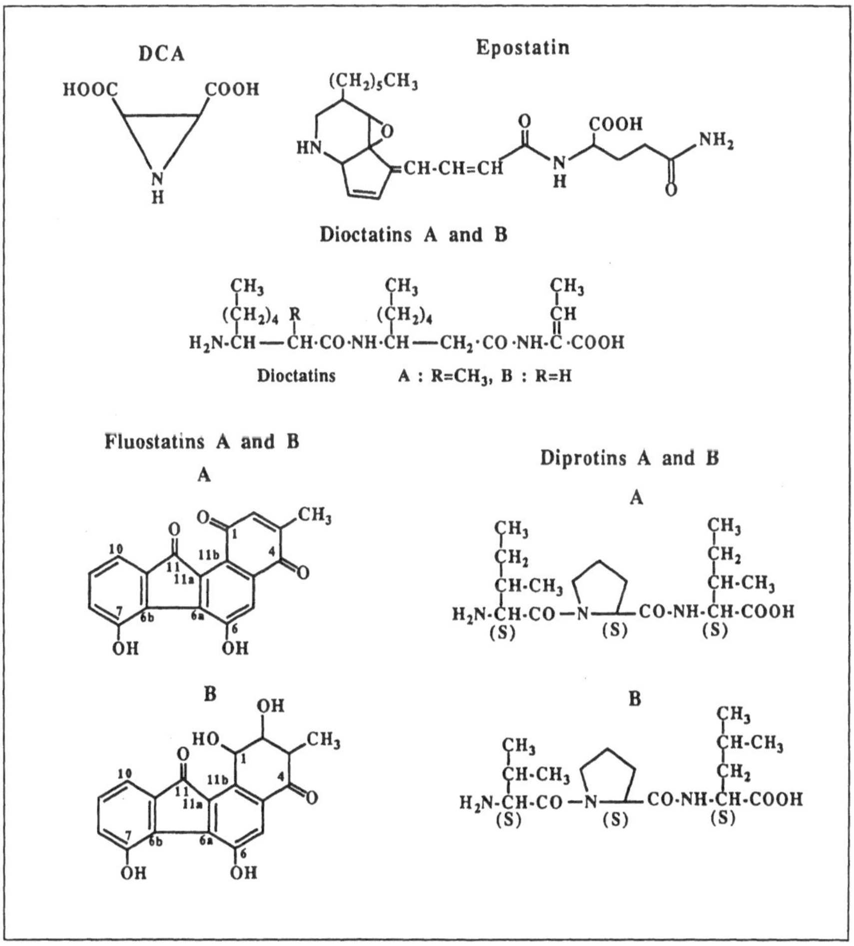

Fig. 7.2. Structures of dipeptidylpeptidase inhibitors. DCA = 2,3-Dicarboxyaziridine.

ROLE OF ECTOENZYMES AND THEIR INHIBITORS IN PATHOLOGIC CONDITIONS

IMMUNOMODULATORY AND THERAPEUTIC CHARACTERISTICS OF BESTATIN (UBENIMEX)

Many reports have been published on the immunomodulatory action of bestatin. Bestatin augmented, in a dose-dependent manner, blastogenesis induced in the presence of mitogen such as Con A, PHA, or LPS.[48-50] A similar augmentation was demonstrated in a mixed culture of lymphocytes.[51] On immunization of C3H/He mice with a vaccine that was prepared by x-irradiation of syngeneic tumor X5563, induction of cytotoxic T cells against the tumor cells was enhanced if bestatin was added to the vaccine.[51] When DNA-synthesizing activities in various organs after in vivo administration of bestatin were examined, it was demonstrated that DNA synthesis in T cells of spleen and thymus, as well as in bone marrow cells, was remarkably enhanced, and DNA polymerization, an activity in T cells, was increased.[52] It is reasonable to assume that the administration of bestatin potentiates the ability of T cells and bone marrow cells to divide and/or to proliferate. Augmentation of blastogenesis demonstrated in vitro can explain the above data obtained with whole animals. Furthermore, it was demonstrated that bestatin activates peritoneal and alveolar macrophages; it also recovers the suppressed activity of NK cells induced by tumor bearing.[53]

Antitumor activity of bestatin, as demonstrated using various animal models, is summarized in Table 7.6. Bestatin showed antitumor activity on IMC carcinoma, myeloid leukemia C 1498, and colon adenocarcinoma 26 implanted to mice subdermally.[54,55] In a model experiment to examine the effect on metastasis of leukemia P388 cells to axillary lymph node, the inhibitor suppressed not only the metastasis but also the development of minor oncogenesis induced by the metastasis.[56] Furthermore, bestatin inhibited the growth of gastrocarcinoma induced by a chemical carcinogen, N-methyl-N-nitro-N-nitrosoguanidine.[57] Clinically, bestatin is now widely employed to remove a small number of tumor cells remaining after chemotherapeutic treatment with some cytotoxic antitumor agents. It also prevents metastasis of a few residual cells to lymph nodes after surgical treatment.

In addition, bestatin is a useful reagent in immunology.[58] It affects cellular receptors toward growth-differentiation factors such as CSF (colony stimulating factor), macrophages, and T cells, some through its indirect action through production of a series of cyto-kinins.

Table 7.5. Kinetic constants of dipeptidylpeptidase inhibitors

Inhibitor	Enzyme	Substrate	Km $(\times 10^{-4}M)$	K_i $(\times 10^{-7}M)$	Type of inhibition
DCA[1]	DPP-I	Gly-Arg·NA	8.0	42.0	Competitive
Epostatin	DPP-II	Lys-Ala·NA	14.3	14.4	Competitive
Fluostatin A	DPP-III	Arg-Arg·NA	0.6	6.59	Mixed type
Fluostatin A	DPP-III	Leu-enkephalin	0.1	142.0	Competitive
Diprotin A	DPP-IV	Gly-Pro·NA	4.0	22.0	Competitive
Diprotin B	DPP-IV	Gly-Pro·NA	4.0	76.0	Competitive

(1) 2,3-Dicarboxyaziridine

Table 7.6. Antitumor-associated biological activities of bestatin in experimental models

Activity	Animal	Tumor
Growth inhibition	Mice	IMC carcinoma, Myeloid leukemia C1498 Colon adenocarcinoma 26, MCA fibrosarcoma Gardner's lymphosarcoma, Sarcoma 180
	Rats	FMT fibrosarcoma
Increase of survival time	Mice	Ascites hepatoma MH-134
	Rats	Ascites hepatoma AH-44, 66, 130
Suppression of carcinogenesis	Mice	Skin cancer (by 20-methylcholanthrene) Spontaneous age-related tumor
	Rats	Nitrosoguanidine-induced stomach tumor
Inhibition of metastasis		P388 Leukemia (lymph node) Lewis lung carcinoma, B16 melanoma BL-6 (lung) L5178Y Esb leukemia (liver)
Combination with chemotherapy		Daunorubicin, Cyclophosphamide (L1210 leukemia) Mitomycin C, 5-Fluorouracil, Cisplatin (colon adenocarcinoma 26) Bleomycin (Ehrlich carcinoma, AH-66 hepatoma)

EFFECT OF BESTATIN (UBENIMEX) IN TUMOR CELL, INVASION AND METASTASIS

Malignant tumors are characterized by an unlimited proliferation or division of somatic cells. Metastasis of the malignant cells, i.e., dissemination of the malignant cells or the focus from the original site of oncogenesis into remote organs, is the most critical factor in determining the mortality of the disease. Metastasis and penetration into organs, lymph glands, or blood vessels involves altered mobility, adhesion, and altered ability to destroy extracellular matrix.[59,60] Extracellular proteases and glycosidases appear to be involved in destroying extracellular matrix. Enzymes which are produced by the host cells themselves also take part in the process.[61]

Bestatin shows immunomodulatory action and is now widely employed clinically as an antitumor agent with host-mediated actions. It suppresses metastasis of mouse P388 leukemia cells to lymph nodes[62] and shows therapeutic effects on natural and experimental metastasis of B16-BL6 melanoma if a suitable dose is used.[63] Recently, it was shown that the effect is due to the direct inhibition of AP-N by the inhibitor.[64,65]

Matrix metalloproteases (MMP), which play an important role in tumor invasion and metastasis, have Zn^{2+} in their active center. The primary structure around that portion is similar to that of AP-N. Actinonin, an AP-N inhibitor, also inhibits MMP.

PHARMACOLOGIC APPLICATIONS OF ENZYME INHIBITORS

The biological and therapeutic effects of enzyme inhibitors that we investigated are summarized (Chart 7.1). Immunopotentiation activity was found for actinonin, amastatin, and bestatin. Fertilization is known to be suppressed by antipain, chymostatin, and leupeptin. Bestatin and arphamenines A and B have analgesic activity. Carageenin-induced edema is suppressed by antipain, chymostatin, and elastatinal, and ascites and pleural fluid volumes are suppressed by pepstatin. This is consistent with suppressed burn-induced, blister formation by leupeptin. Leupeptin also inhibits the release of serine proteinases in pancreatitis patients. The high-plasma renin activity in renal vascular hypertension is inhibited by pepstatin. Ebelactone B has been associated with diminished hyperlipemia. Suppression of some malignancies is suggested

Chart 7.1. Proposed biological and therapeutic effects of enzyme inhibitors

Immunomodulator	Ac, Am, Ar(A,B), Bm, Bs, Dp(A,B), Eb(A,B), Es, Fl, Fn, Fr, Hs, Le, Fo, Pr
Fertilization	Ap, Cs, Lp
Analgesic	Ac, Am, Ar(A,B), Bs
Inflammation	Ap, Cs, El, Es, Fr, Lp, Ph, Ps
Ascites and pleural fluid	Ps
Burn	Lp
Pancreatitis	Lp
Hypertension	Bs, Fo, Hs, Ps
Hyperlipemia	Eb(B)
Malignant diseases	Bs, Fl, Fn
Autoimmune diseases	Bs, Fl, Fn, Po
Alzheimer's diseases	Po

Ac, actinonin; Am, amastatin; Ap, antipain; Ar, arphamenine; Bm, (S)-a-benzylmalic acid; Bs, bestatin; Cs, chymostatin; Di, dioctatin; Dp, diprotin; Eb, ebelactone; El, elastatinal; Es, esterastin; Fl, forphenicinol; Fn, forphenicine; Fo, foroxymithine; Fr, formestin; Hs, histargin; Le, leuhistin; Lp, leupeptin; Ph, phosphoramidon; Po, poststatin; Pr, probestin; Ps, pepstatin

for bestatin, forphenicine, and forphenicinol. Autoimmune diseases, including demyelinative diseases, appear to be suppressed by bestatin, torphenicinol and poststatin. Therapeutic effects on Alzheimer's disease was suggested for poststatin.

SUMMARY

In living organisms a large number of enzymes are working in complicated networks to express various biological functions. In order to analyze such functions from various aspects, specific enzyme inhibitors are likely to become useful tools. Searching for inhibitors in culture filtrates of microbes, we discovered many specific inhibitors of various enzymes, such as endopeptidases, exopeptidases, glycosidases and lipases, some of which are on cell surfaces. These inhibitors have low molecular weights and unique structures. They are also useful for studies of the reaction mechanisms and analysis of three-dimensional structures of enzymes[66-68] (see chapter 2). They are also of value in elucidating disease processes and seem to have usefulness in the treatment of various diseases. Some of the inhibitors are now being tested in clinical studies.

REFERENCES

1. Umezawa H. Enzyme Inhibitors of Microbial Origin. University of Tokyo Prees, Tokyo 1972:1-114.
2. Aoyagi T, Umezawa H. Structures and activities of protease inhibitors of microbial origin. In: Reich E, Rifkin D, Shaw E, eds. Proteaseases and Biological Control. Cold Spring Harbor Laboratory, 1975:429-454.
3. Aoyagi T, Ishizuka M, Takeuchi T et al. Enzyme inhibitors in relation to cancer therapy. Jap J Antibiotics (Suppl) 1977; 30-S:121-132.
4. Aoyagi T. Bioactive peptides produced by microorganisms. In: Umezawa H, Shiba T, Takita T, eds. Kodansha Scientific Press, 1978:129-151.
5. Aoyagi T, Umezawa H. Industrial and clinical enzymology. In: Vitale L, Simeon V, eds. FEBS Federation of European Biochemical Societies. 61. Oxford and New York: Pergamon Press, 1980:89-91.
6. Umezawa H. Annual review of microbiology. In: Starr MP, Balows A, Schmidt JM, eds. Vol. 36. Palo Alto: Annual Reviews Inc., 1982:75-99.
7. Aoyagi T. Horizons on antibiotic research. In: Davis BD, Ichikawa T, Maeda K, Mitcher LA, eds. Tokyo: Japan Antibiotics Research Association, 1987:75-90.
8. Aoyagi T. Progress in Industrial Microbiology. Bushell ME, Grafe U, eds. Vol. 27. New York, Oxford: Elsevier, 1989:403-418.
9. Aoyagi T. Small molecular protease inhibitors and their biological effects. In: Kleinkauf H, Dohren H, eds. Biochemistry of Peptide Antibiotics. Berlin, New York: Walter de Gruyter, 1990:311-363.
10. Aoyagi T, Nerome K, Suzuki J et al. Change of enzyme activities during the early stage of influenza virus infection. Biochem Biophys Res Commun 1974; 60:1178-1184.
11. Aoyagi T, Suda H, Nagai M et al. Aminopeptidase activities on the surface of mammalian cells. Biochim Biophys Acta 1976; 452:131-143.
12. Aoyagi T, Nagai M, Iwabuchi M et al. Aminopeptidase activities on the surface of mammalian cells. Cancer Res 1978; 38:3505-3508.
13. Kalckar HM. Science 1965; 150:305-308.
14. Aoyagi T, Suzuki J, Nerome K et al. Sialic acid residues exposed on mammalian cell surface: the effect of adsorption of denatured virus particles. Biochem Biophys Res Commun 1974; 57:271-278.
15. Aoyagi T, Komiyama T, Nerome K et al. Characterization of myxovirus sialidase. Experientia 1975; 31:896-897.
16. Aoyagi T, Umezawa H. Hydrolytic enzymes on the cellular surface and their inhibitors found in microorganisms. In: Vitale L, Simon V, eds. Industrial and Clinical Enzymology. Oxford and New York: Pergamon Press, 1980:89-99.

17. Aoyagi T, Umezawa H. The relationships between enzyme inhibitors and function of mammalian cells. Acta Biol Med Germ 1981; 40:1523-1529.
18. Fujii H, Nakajima I, Tsuruo T. Role of aminopeptidases on metastasis. Hematology & Oncology (in Japanese) 1994; 29:288-296.
19. Sanderink GJ, Artur Y, Siest G. Human aminopeptidases: a review of the literature. J Clin Chem Clin Biochem 1988; 26:795-805.
20. Kenny AJ, Stephanson SL, Turner AJ. Research monographs in cell and tissue physiology. In: Mammalin Ectoenzyme. Amsterdam: Elsevier, 1987:169-210.
21. Lalu K, Lampelo S, Vanha-Perttula T. Characterization of three aminopeptidases purified from maternal serum. Biochim Biophys Acta 1986; 873:190-198.
22. Look AT, Ashmun RA, Shapiro LH et al. Human mycloid plasma membrane glycoprotein CD13 (gp150) is identical to aminopeptidase. J Clin Invest 1989; 83:1299-1307.
23. Sanderink GJ, Artur Y, Schiele F et al. Alanine aminopeptidase in serum: biological variations and reference limits. Clin Chem 1988; 34:1422-1430.
24. Hersh LB. Characterization of membrane-bound aminopeptidases from rat brain: identification of the enkephalin-degrading aminopeptidase. J Neurochem 1985; 44:1427-1435.
25. Bausback HH, Ward PE. Degradation of low-molecular-weight opioid peptides by vascular plasma membrane aminopeptidase M. Biochim Biophys Acta 1986; 882:437-442.
26. Amoscato AA, Balasubramaniam A, Alexander JW et al. Degradation of thymopentin by human lymphocytes: evidence for aminopeptidase activity. Biochim Biophys Acta 1988; 955:164-174.
27. Aoyagi T, Tobe H, Kojima F et al. Amastatin, an inhibitor of aminopeptidase A, produced by actinomycetes. J Antibiotics 1978; 31:636-638.
28. Umezawa H, Aoyagi T, Suda H et al. Bestatin, an inhibitor of aminopeptidase B, produced by actinomycetes. J Antibiotics 1976; 29:97-99.
29. Nishizawa R, Saino T, Takita T et al. Synthesis and structure-activity relationships of bestatin analogues, inhibitors of aminopeptidase B. J Med Chem 1977; 20:510-515.
30. Umezawa H, Ishizuka M, Aoyagi T et al. Enhancement of delayed-type hypersensitivity by bestatin, an inhibitor of aminopeptidase B and leucine aminopeptidase. J Antibiotics 1976; 29:857-859.
31. Umezawa H, Aoyagi T, Ohuchi S et al. Arphamenine A and B, new inhibitors of aminopeptidase B, produced by bacteria. J. Antibiotics 1983; 36:1572-1575.
32. Ohuchi S, Suda H, Naganawa H et al. The structure of arphamenine A and B. J Antibiotics 1983; 36:1576-1580.
33. Okuyama A, Ohuchi S, Tanaka T et al. Cell-free biosynthesis of

arphamenine A. Biochem Int 1986; 12:485-491.

34. Yamamoto K, Suda H, Ishizuak M et al. Isolation of *a*-aminoacyl arginines in screening of aminopeptidase B inhibitors. J Antibiotics 1980; 33:1597-1599.

35. Look TA, Ashmun RA, Shapiro LH et al. Human myeloid plasma membrane glycoprotein CD13 (gp150) in identical to aminopeptidase N. J Clin Invest 1989; 83:1299-1307.

36. Aoyagi T, Yoshida S, Nakamura Y et al. Probestin, a new inhibitor of aminopeptidase M, produced by *Streptomyces azureus* MH663-2F6. I. Taxonomy, production, isolation, physico-chemical properties and biological activities. J Antibiotics 1990; 43:143-148.

37. Yoshida S, Nakamura Y, Naganawa H et al. Probestin, a new inhibitor of aminopeptidase M, produced by *Streptomyes azureus* MH663-2F6. II. Structure determination of probestin. J Antibiotics 1990; 43:149-153.

38. Aoyagi T, Yoshida S, Matsuda N et al. Leuhistin, a new inhibitor of aminopeptidase M, produced by *Bacillus laterosporus* BMI156-14F1. I. Taxonomy, production, isolation, physico-chemcial properties and biological activities. J Antibiotics 1991; 44:573-578.

39. Yoshida S, Naganawa H, Aoyagi T et al. Leuhistin, a new inhibitor of aminopeptidase M, produced by *Bacillus laterosporus* BMI156-14F1. II. Structure determination of leuhistin. J Antibiotics 1991; 38:579-581.

40. Umezawa H, Aoyagi T, Tanaka T et al. Production of actinonin, an inhibitor of aminopeptidase M, by actinomycetes. J Antibiotics 1985; 38:1629-1630.

41. Umezawa H, Aoyagi T, Uotani K et al. Ebelactone, an inhibitor of esterase, produced by actinomycetes. J Antibiotics 1980; 33:1594-1596.

42. Uotani K, Naganawa H, Kondo S et al. Structure studies on ebelactones A and B, esterase inhibitors produced by actinomycetes. J Antibiotics 1982; 35:1495-1499.

43. Umezawa H, Aoyagi T, Hazato T et al. Esterastin, an inhibitor of esterase, produced by actinomycetes. J Antibiotics 1978; 31:639-641.

44. Hegen M, Niedobitek G, Klein CE et al. The T cell triggering molecule Tp103 is associated with dipeptidyl aminopeptidase IV activity. J Immunol 1990; 144:2908-2914.

45. Ulmer AJ, Mattern T, Feller AC et al. CD26 antigen is a surface dipeptidyl peptidase IC (DPP IV) as characterized by monoclonal antibodies done TII-19-4-7 and 4ELIC7. Scand J Immunol 1990; 31:429-435.

46. Torimoto Y, Dang NH, Tanaka T et al. Biochemical characterization of CD26 (dipeptidyl peptidase IV): functional comparison of direct epitopes recognized by various anti-CD26 monoclonal antibodies. Mol Immunol 1992; 29:183-192.

47. Umezawa H, Aoyagi T, Ogawa K et al. Diprotins A and B, inhibitors of dipeptidyl aminopeptidase IV, produced by bacteria. J Antibiotics 1984; 37:422-425.

48. Ishizuka M, Sato J, Sugiyama Y et al. Mitogenic effect of bestatin on lymphocytes. J Antibiotics 1980; 33:653-662.

49. Naito M, Aoyagi T, Umezawa H et al. Bestatin, a new specific inhibitor of aminopeptidases, enhances activation of small lymphocytes by concanavalin. Biochem Biophys Res Commun 1977; 76:525-533.

50. Abe F, Kuramochi H, Takahashi K et al. Biological activity of the main metabolites of ubenimex in humans. J Antibiotics 1988; 41:1862-1867.

51. Talmadge JE, Lenz BF, Pennington R et al. Immunomodulatory and therapeutic properties of bestatin in mice. Cancer Res 1986; 46:4505-4510.

52. Muller WEG, Zahn RK, Arendes J et al. Activation of DNA metabolism in T-cells by bestatin. Biochem Pharmacol 1979; 28:3131-3137.

53. Talmadge JE, Koyama M, Matsuda A et al. Immunotherapeutic properties of bestatin: mechanism of activity. In: Umezawa H, ed. Recent Results of Bestatin 1986-A biological response modifier. Tokyo: Jpn Antibiot Res Associ, 1986:8-25.

54. Ishizuka M, Masuda T, Kanbayashi N et al. Effect of bestatin on mouse immune system and experimental murine tumors. J Antibiotics 1980; 33:642-652.

55. Abe F, Shibuya K, Uchida M et al. Effect of bestatin on syngeneic tumors in mice. Gann 1984; 75:89-94.

56. Tsuruo T, Naganuma H, Iida H et al. Inhibition of lymph node metastasis of P388 leukemia by bestatin. J Antibiotics 1981; 34:1206-1209.

57. Ebihara K, Abe F, Yamashita T et al. The effect of ubenimex on N-methyl-N'-nitro-N-nitrosoguanidine-induced stomach tumor in rats. J Antibiotics 1986; 39:966-970.

58. Horiuchi K, Miyamoto T et al. Radioprotective effect of ubenimex (bestatin) on C3H and BALB/c mice. 16th ICC proceedings, Jerusalem, 1989.

59. Fidler IJ, Fabra A, Nakajima M et al. Genetic and Epigenetic regulation of human colon carcinoma metastasis. Contrib Oncol 1992; 44:13-21.

60. Nicolson GL. Paracrine and autocrine growth mechanisms in tumor metastasis specific sites with particular emphasis on brain and lung metastasis. Cancer Metastasis Rev 1993; 12:325-338.

61. Nakajima M, Chop AM. Tumor invasion and extracellular matrix degrading enzyme: Regulation of activity by organ factors. Semin Cancer Biol 1991; 2:115-123.

62. Tsuruo T, Naganuma K, Iida H et al. Inhibition of lymph node

metastasis of P388 leukemia by bestatin in mice. J Antibiotics 1981; 34:1206-1209.

63. Talmadge JE, Lenz BF, Pennington R et al. Immunomodulatory and therapeutic properties of bestatin in mice. Cancer Res 1986; 46:4505-4510.

64. Saiki I, Murata J, Watanabe K et al. Inhibition of tumor cell invasion by ubenimex (bestatin) in vitro. Jap J Cancer Res 1989; 80:873-878.

65. Saiki I, Fujii H, Yoneda J et al. Role of aminopeptidase N (CD13) in tumor-cell invasion and extracellular matrix degradation. Int J Cancer 1993; 54:137-143.

66. Taylor A, Sawan S, James T. On the binding of leucyl-o-sulfonic acid in leucine aminopeptidase. Interaction between this substrate analog and the activation site metal-viewed by NMR. J Biol Chem 1982; 257:11571-6.

67. Taylor A, Peltier CZ, Torre FJ et al. Inhibition of bovine lens leucine aminopeptidases by bestatin: number of binding sites and slow binding of this inhibitor. Biochemistry 1993; 32:784-90.

68. Taylor A, Peltier CZ, Jahngen EGE et al. Use of azidobestatin as a photoaffinity label to identify the active site peptide of leucine aminopeptidase. Biochemistry 1992; 31:4141-4150.

PLANT AMINOPEPTIDASES: OCCURRENCE, FUNCTION AND CHARACTERIZATION

Linda L. Walling and Yong-Qiang Gu

INTRODUCTION

Peptidases have a central role in the degradation of proteins by hydrolyzing peptide bonds. Based on the reaction catalyzed, these enzymes can be classified as endopeptidases, aminopeptidases or carboxypeptidases. Relative to the endoproteases and carboxypeptidases, the three types of exopeptidases that act at free N-termini of proteins or peptides are less well characterized. Aminopeptidases, dipeptidyl peptidases and tripeptidyl peptidases cleave peptides or proteins and liberate single amino acid residues, dipeptides or tripeptides, respectively. Dipeptidases and tripeptidases are aminopeptidases that preferentially act on dipeptides and tripeptides. While all three classes of N-terminal exopeptidases have been described in animals, only aminopeptidases have been described in plants.

Aminopeptidases are ubiquitous enzymes, and a wide variety of aminopeptidase activities have been detected in animal, plant and prokaryotic cells. While the biological functions of some aminopeptidases are still unknown, others are well documented. For example, methionine aminopeptidases remove N-terminal methionines and are critical for the maturation of many proteins in

---

Aminopeptidases, edited by Allen Taylor. © 1996 R.G. Landes Company.

eukaryotic and prokaryotic cells. Deletion of the gene encoding the methionine aminopeptidase is a lethal event for *Salmonella typhimurium* and significantly retards cell growth in *Saccharomyces cerevisiae*.[1,2] It is well established that aminopeptidases are important in the activation and inactivation of some regulatory molecules, including peptide hormones and neurotransmitters, and modulation of cell-cell interactions.[3,4] For example, the membrane-associated aminopeptidases A and N act sequentially to inactivate the peptides angiotensin II and III, and soluble and membrane-bound aminopeptidases act in concert to process the α-bag peptide hormones of *Aplysia*.[5,6] Aminopeptidases may also play an active role in regulation of protein turnover. Since aminopeptidases hydrolyze the N-terminal amino acids from proteins and peptides at different rates depending on their substrate specificities and N-terminal residues vastly influence protein half-lives by deterring or potentiating ubiquitination, aminopeptidases may target proteins for degradation by the 20S proteosome.[7-13]

In addition to their exopeptidase activities, some aminopeptidases have second biological functions. For example, aminopeptidase N is a receptor for human and mouse coronaviruses, and the leucine aminopeptidase analog of *Escherichia coli* (XerB) facilitates site-specific recombination at the *cer* site of ColE1 plasmids.[14-16] In both cases, the aminopeptidase activity domains are distinct from the domains involved in recombination or viral attachment.[17-18]

In plants, aminopeptidases are less well characterized. Aminopeptidases were first detected in higher plants over 60 years ago.[19] In the 1970s and 1980s, a large number of studies identified plant enzymes that hydrolyzed amino acyl-β-naphthylamides, amino acyl-p-nitroanilides and/or peptides efficiently.[20] These research initiatives showed the diversity and ubiquity of the aminopeptidase activities present in the plant kingdom. In some cases, these hydrolytic activities were detected in tissues where accelerated protein turnover was documented. Aminopeptidase activities were detected after monocot and dicot seed germination and during leaf, petal and ovary senescence.[21-25] Although their physiological role has not been rigorously tested, it is believed that these exopeptidases may facilitate protein degradation to mobilize storage proteins or salvage carbon and nitrogen from cells committed to death. Exoproteases were also detected in tissues undergoing rapid cell divisions; high levels of aminopeptidases were found during early stages of leaf and seedling development, and during seed or fruit maturation.[26-30] Finally, the recent findings that the plant peptide hormone systemin is a wound signal, N- and C-terminal processing is essential for the maturation of this peptide hormone, and that aminopeptidases are induced during the plant defense response, have given the field of plant protein processing and turnover a new vitality.[31-35]

It has been ten years since the last overview of plant aminopeptidases. Since that time, the first two classes of plant aminopeptidase clones have been isolated and show significant similarity to analogs in mammals. The barley seed peptidase aleurain has numerous characters that indicate it is an aminopeptidase and is analogous to the human cathepsin H.[36] In addition, cDNAs from *Arabidopsis*, potato and tomato form the first unified plant aminopeptidase family which has similarity to the well-characterized bovine lens leucine aminopeptidase (LAP).[34,35,37] The classification schemes for the remaining plant aminopeptidases that have been characterized will be discussed, and the possible physiological functions of these aminopeptidases during plant development will be presented. The role of aminopeptidases in the defense response and in response to environmental stress will also be addressed.

CLASSIFICATION OF PLANT AMINOPEPTIDASES

The grouping of endopeptidases and carboxypeptidases based on the structure of catalytic sites has allowed for the development of a simple and useful scheme of classification.[38] It has been used as a framework to discuss the evolutionary relatedness, structural similarities, and in some cases the functional homology of endopeptidases and carboxypeptidases.[39] As noted in recent reviews, a similar system of classification for aminopeptidases is lacking.[40-41] Although a small number of aminopeptidases and dideptidyl peptidases are well characterized at the molecular and/or structural level, there is still an inadequate number of prototype aminopeptidases that can be used to establish this essential comparative framework. It is imperative that such a system be developed in the near future to assure that structural and functional similarities and differences of aminopeptidases are fully recognized.

Traditionally, aminopeptidases were classified solely based on the substrate used in an activity assay. This resulted in aminopeptidases being given names that are not reflective of their substrate specificities, many distinct aminopeptidases being given similar or the same names, and the identical exopeptidases being given different names.[40,42-45] The classification system for enzymes called leucine aminopeptidase (LAP) is particularly complex. The best characterized aminopeptidase in the scientific literature is the hexameric leucine aminopeptidase from bovine lens.[40,41] Not only is there substantial literature characterizing the animal LAP at the biochemical level, the three-dimensional structure at the 2.7 Å level is known, and plant and prokaryotic homologs have been identified.[7,16,34,35,37,46-48] This hexameric LAP is structurally and biochemically distinct from the monomeric aminopeptidases that bear the same name and hydrolyze leucine-p-nitroanilide or leucine-β-naphthylamide substrates efficiently. These enzymes were originally designated as leucine aminopeptidases and have been extensively used as isozyme markers in plant genetic studies.[45] In the strictest sense, these monomeric exopeptidases should be designated as naphthylamidases or arylamidases until peptide processing capability is established. For ease of discussion, the plant naphthylamidases and arylamidases will be called aminopeptidases.

Although a large number of aminopeptidase activities have been identified in crude extracts of plant tissues, a smaller number have been partially purified to either enzymatic or biochemical homogeneity (Tables 8.1-8.4). To date, only four plant aminopeptidase genes have been cloned.[34-37] While the plant leucine aminopeptidases (LAPs) form a discrete class of plant aminopeptidases based on nucleic acid and peptide sequence conservation and structural and biochemical properties, it is evident that the remaining exopeptidases represent a diverse array of biochemical activities. The system of classification for plant aminopeptidases proposed by Mikola and Mikola in 1986 still serves as a useful guide for discussing plant exopeptidases. Using optimal pH for enzymatic activity and substrate specificities as criterion, they identified two major aminopeptidase divisions: the neutral aminopeptidases and the alkaline aminopeptidases. Although it has been ten years since the Mikola and Mikola classification system was proposed, aminopeptidases with acidic pH optima have yet to be identified, although an array of acidic endopeptidases and carboxypeptidases that are localized in the vacuole have been described in plants.[21-22]

Table 8.1. *Plant neutral aminopeptidases that prefer aromatic and hydrophobic residues*[A]

Plant species	Name	MW (kDa)	Substrates[B]	pH Optimum	Inhibitors	Metallo-enzyme[C]	Active on peptides?	Ref.
Euonymous	LAPase2	62.5	Leu-, Phe-, Tyr-β-NAP	7.6	pCMB, Cu^{+2}, Hg^{+2}, Zn^{+2}	no	n.d.	53
Glycine max (soybean)	AP	76-85	Trp-, Phe-, Tyr-, Leu-,Met-β-NAP	n.d.	pCMB, Zn^{+2}, Cu^{+2}, Ag^{+2}	no	yes	54
Hordeum vulgare (barley)	AP	65	Phe-, Tyr-, Leu-, Met-β-NAP	7.2	pCMB	no	yes	55
Pisum sativum (pea)	AP1	58	Phe-, Leu-, Tyr-β-NAP	6.8	pCMB, $Zn^{+2}Cu^{+2}$, Hg^{+2}	no	yes	51,52
Prunus mime (apricot)	AP1	56	Phe-, Leu-, Tyr, Trp-b-NAP	6.5-7.0	pCMB, Cu^{+2}, Hg^{+2}, Zn^{+2}	no	yes	57,58
Triticum aestivum (wheat)	AP1	57	Phe-, Trp-, Tyr-, Leu-β-NAP	7.6	pCMB, $Zn^{+2}Cu^{+2}$, Ag^{+2}	yes[D]	yes	26
Vigna radiata (mung bean)	AP	70-75	Phe-, Leu-, Met-p-NA	n.d.	pCMB, Hg^{+2}	n.d.	yes	58
Vigna unguiculta (cowpea)	AP2	n.d.	Phe-, Leu-, Met-β-NAP	n.d.	Zn^{+2}, Ag^{+2}	no	n.d.	49
Zea mays (maize)	AMP4	61.3	Phe-, Tyr-, Leu-, Trp-β-NAP	6.9	pCMB, $Cu^{+2}Zn^{+2}$, Hg^{+2}	no	n.d.	59

A Throughout this table n.d. = not determined. The *Chlamydomonas* 76 kDa aminoepeptidase has a molecular weight and substrate specifcity simliar to the Phe aminopeptidases except that its pH optimum is slightly alkaline (Table 8.3).
B The amino acyl-p-nitroanilide (p-NA) and -β-naphthylamide (β-NAP) substrates hydrolyzed most efficiently are listed.
C pCMB,p-chloromecurobenzoate
D The wheat AP1 was insensitive to phenanthroline but sensitive to high concentrations of bathocuprine.

Table 8.2. Plant neutral aminopeptidases that prefer alanine or leucine residues[A]

Plant species	Name	MW (kDa)[B]	Substrates[C]	pH Optimum	Inhibitors[D]	Metallo-enzyme	Active on peptides?	Ref.
Agave americana	AP	87.5	Ala-, Lys-, Arg-, Leu-β-NAP	7.2	DEP	n.d.	yes	66, 67
Chara australis	API	120 (44 & 62)	Ala-, Phe-, Arg-, Lys-b-NAP	6.0-7.5	pCMB	yes	n.d.	60
	APII	85	Ala-β-NAP	7.0	pCMB	yes	n.d.	61
Cucurbita maxima (squash)	tripep-tidase	n.d.	Leu-Gly-Gly Gly-Gly-Gly	7.0	n.d.	yes	yes	62
Euglena gracilis	LeuAP1	390	Leu-, Ala-β-NAP	7.2	n.d.	n.d.	n.d.	68
Ipomea batatas (sweet potato)	AP1	115	Ala-, Leu-β-NAP	7.0	pCMB, PMSF	yes	yes	63, 64
	AP2	63	Leu-, Ala-β-NAP	7.6	pCMS, PMSF, Zn^{+2}	yes	yes	
Simmondsia chinensis (jojoba)	AP	14.2	Leu-p-NA; Leu-β-NAP	6.9	pCMB	no	n.d.	70
Oryza sativa (rice)	AP	n.d.	Leu-p-NA	7.2	pCMB	yes	yes	65
Pisum sativum (pea)	AP2	74	Ala-, Arg-, Leu- -NAP	7.0	pCMB	yes	n.d.	51
Vigna unguiculta (cowpea)	AP1	n.d.	Arg-, Ala-, Lys-β-NAP	n.d.	Zn^{+2}, Ag^{+2}	yes	n.d.	49
	AP3	n.d.	Ala-, Gly-b-NAP	n.d.	Zn^{+2}, Ag^{+2}	yes	n.d.	49
Vitis vinifera (grape)	AP	95 (33 & 62)	Ala-β-NAP Ala, Leu-pNA	n.d.	pCMB, NEM	n.d.	yes	69

A Throughout this table, n.d. = not determined. The *Chlamydomonas* 92 kDa Ala AP has a substrate specificity similar to the Ala aminopeptidases in this table, however its pH optimum is slightly alkaline (7.7) and is therefore listed in Table 8.3.

B Molecular weight of native enzyme is expressed in kDa. The sizes of the aminopeptidase subunits are indicated in parentheses.

C The major amino acyl-p-nitroanilide (pNA), amino acyl-β-naphthylamide (β-NAP) or peptide substrates for the aminopeptidase are listed.

D pCMB, p-Chloromecurobenzoate; NEM, N-ethylmaleimide; DEP, diethylpyrocarbonate; PMSF, phenylmethylsulfonylfluoride.

Table 8.3. Plant alkaline aminopeptidases[A]

Plant species	Name	MW (kDa)[B]	Substrates[C]	pH Optimum	Inhibitors[D]	Metallo-enzyme	Active on peptides?	Ref.
Euglena gracilis	Ala-AP1	68	Ala-, Gly-β-NAP	8.5	n.d.	n.d.	n.d.	68
	Ala-AP2	68	Ala-, Gly-β-NAP	9.0	n.d.	n.d.	n.d.	68
Cucurbita maxima (squash)	AP	n.d.	Gly-Gly-Gly, Leu-Gly-Gly	8.0		yes	yes	91
Hordeum vulgare (barley)	dipeptidase	130-175 (50)	Ala-Gly	8.8	pCMB	yes	yes	94
Pinus sylvestris (Scots pine)	AP	n.d.	Ala-Gly	7.8	n.d.	n.d.	yes	73
Pisum sativum (pea)	AP2	n.d.	Ala, Gly, Lys-β-NAP	8.0	pHMB, Cu^{+2}, bestatin, Zn^{+2}	yes	n.d.	75
Phaseolus vulgaris (kidney bean)	dipeptidase	105	Ala-Gly, Ala-Leu, Ala-Val	8.5	pCMB, Zn^{+2}, bestatin	yes	yes	87
Zea mays (corn)	AMP2	86.5	Ala-, Gly-, Met-β-NAP	8.5	pCMB, Zn^{+2}, Cu^{+2}, Hg^{+2}	no	yes	59
Avena sativa (oat)	AP	65	Leu-p-NA	8.4	Zn^{+2}	n.d.	n.d.	92
Cucurbita maxima (squash)	dipeptidase	n.d.	Leu-Gly	8.0-8.5	Zn^{+2}, Co^{+2}	yes	yes	96

Table 8.3. *Plant alkaline aminopeptidases[A] (continued)*

Plant species	Name	MW (kDa)[B]	Substrates[C]	pH Optimum	Inhibitors[D]	Metallo-enzyme	Active on peptides?	Ref.
Pinus sylvestris (Scots pine)	AP	n.d.	Leu-Tyr	8.6	n.d.	n.d.	yes	73
Haynaldotium sardoum	AP	89	Leu-, Met-p-NA	8.0	pCMB, pHMB, bestatin	n.d.	n.d.	97
Hordeum vulgare (barley)	AP3	54.6	Arg--NAP	n.d.	pCMB	yes	n.d.	23
Zea mays (corn)	AMP3	83	Lys-, Arg-, Met-, Leu-β-NAP	8.1	pCMB, Zn⁺², Cu⁺², Hg⁺²	yes	yes	59
Chlamydomonas reinhardii	AP	92	Ala-, Lys-, Leu-pNA	8.0-8.5	Zn⁺²	n.d.	n.d.	82
	AP	76	Phe-, Leu-, Tyr-pNA	8.0-8.5	n.d.	n.d.	n.d.	82
	IP	225	Pro-pNA	8.0-8.5	n.d.	n.d.	n.d.	82

A Throughout this table n.d. = not determined.
B Molecular weight of native enzyme is expressed in kDa. The sizes of the aminopeptidase subunits are indicated in parentheses; all other alkaline aminopeptidases are thought to be monomers.
C Major amino acyl-p-nitroanilide (pNA), amino acyl-β-naphthylamide (β-NAP) or peptide substrates for the aminopeptidase. The barley and kidney bean dipeptidases do not cleave tripeptides, -p-nitroanalide or -β-naphthylamide substrates.
D pCMB, p-chloromecurobenzoate; pHMB, phydroxymethylbenzoate.

Table 8.4. Plant leucine aminopeptidases[A]

Plant species	Name	MW (kDa)[B]	Substrates[C]	pH Optimum	Inhibitors[D]	Ref.
Arabidopsis thaliana	LAP	320 (55)	Leu-p-NA Leu-β-NAP	8.5	pCMB, Zn^{+2}, EDTA 1,10-phenanthroline	37, 108
Hordeum vulgare (barley)	LAP	260 (n.d.)	Leu-Gly-Gly Met-Gly-Gly Leu-Tyr, Leu-Gly	8.5-10.5	-SH needed for full functionality	95
Phaseolus vulgaris (kidney bean)	LAP	360 (58 & 60)	Leu-Gly-Gly Leu-Tyr, Leu-Gly	9.0	pCMB, bestatin 1,10-phenanthroline	88
Lycopersicum esculentum (tomato)	LAP	> 327 (55)	Leu- -NAP Leu-p-NA Leucinamide	n.d.	n.d.	34
Solanum tuberosum (potato)	LAP	n.d.	Arg-, Leu- Met-p-NA	10.0	EDTA, bestatin, Zn^{+2}	35, 111

A Throughout this table, n.d. = not determined.
B Molecular weight of native enzyme is expressed in kDa. The sizes of the LAP subunits are indicated in parentheses.
C Leu-pNA = Leucine-p-nitroanilide; Leu-β-NAP = Leucine-β-naphthylamide; The barley and kidney bean LAPs have minimal activity on Leu-β-NAP and Leu-pNA when compared to the peptide susbstrates.
D pCMB = p-Chloromecurobenzoate

NEUTRAL AMINOPEPTIDASES

Many of the neutral aminopeptidases were identified as naphthylamidases or arylamidases in zymograms, and this allowed the number of exopeptidases and substrate specificity of each enzyme to be evaluated without enzyme purification.[23,27,45] More than two dozen neutral aminopeptidases have been purified to biochemical or enzymatic purity (Tables 8.1, 8.2). Neutral aminopeptidases can be sorted into several groups based on their substrate specificities including exoproteases that prefer proline, phenylalanine, alanine or basic or acidic residues at their N-termini. Although many aminopeptidases exhibit a substrate preference for selected amino acid residues, some of these exopeptidases also have considerable activity on substrates with N-terminal amino acids with different charges and hydrophobicity.[26,27,49,50]

Nine of the purified neutral aminopeptidases have substrate specificities similar to that displayed by the pea AP1 enzyme that was first described by Elleman (Table 8.1).[26,49,51-59] These "Phe" exopeptidases preferentially hydrolyze β-naphthylamides conjugated to aromatic (Phe, Tyr, Trp) or bulky hydrophobic (Leu, Met) amino acids. All of these exopeptidases are monomers with molecular masses between 56 and 76 kDa and are inhibited by p-chloromercurobenzoate (pCMB) or zinc ions implicating a role for sulfhydryls in the enzyme reaction center or for structural integrity. Only one of the "Phe" aminopeptidases may be a metalloenzyme.[26] All of the "Phe" exopeptidases were first identified as naphthylamidases, and in all cases tested, these enzymes are active on peptide substrates.

A second subgroup of neutral aminopeptidases preferentially hydrolyze substrates with N-terminal alanine or leucine residues (Table 8.2). Nine of the "Ala-Leu" aminopeptidases are metalloenzymes based on their sensitivity to the chelator 1,10-phenanthroline.[49,50,60-65] It is not clear if the grape, *Agave* or *Euglena* aminopeptidases are metalloenzymes, since sensitivity to 1,10-phenanthroline was not tested.[66-69] The "Ala-Leu" aminopeptidases are a collection of enzymes with a wide range of sizes (ranging from 14 to 390 kDa). While many are monomers, the *Chara australia* and grape aminopeptidases are heterodimers, and the subunit composition of the 390 kDa *Euglena* LeuAP1 has yet to be determined. The squash exoprotease is the only plant tripeptidase identified to date; this enzyme hydrolyzes tripeptides and not dipeptides.[62] With the exception of the jojoba aminopeptidase, all "Ala-Leu" aminopeptidases are metalloenzymes.[70] Unlike the other "Ala-Leu" aminopeptidases, the *Chara*, pea, cowpea and *Agave* aminopeptidases also cleave basic substrates efficiently. Most of the "Ala-Leu" aminopeptidases require a sulfhydryl group for activity or to maintain structural integrity; the exception is the *Agave* aminopeptidase that is only inhibited by diethylpyrocarbonate.

Neutral aminopeptidases active primarily on basic amino acid-β-naphthylamide substrates have been detected in crude extracts from leaves and seeds.[27,55,59,71-73] Several alkaline aminopeptidases prefer substrates with N-terminal arginine or lysine residues (Table 8.3), and several "Ala-Leu" aminopeptidases are able to efficiently cleave chromogenic substrates with basic residues (Table 8.2). Only the oat and maize "basic" aminopeptidases have been tested with a broad array of β-naphthylamide substrates. The oat aminopeptidase has strong preference for arginine and lysine residues with low or no activity towards other amino acid residues.[27] In contrast, the maize AMP1 hydrolyzes several substrates efficiently. Only the maize AMP1 basic aminopeptidase was partially purified; it is a metalloenzyme that is inactivated by pCMB.[59]

To date, there is only one example of a plant aminopeptidase that preferentially hydrolyzes acidic amino acids from β-naphthylamide substrates. This enzyme (AP3) was purified from pea seeds and has high affinity towards N-terminal glutamic and aspartic acid residues.[74] AP3 has a native molecular weight of 500 kDa and subunits of 30 and 55 kDa. Like the "Phe" aminopeptidases, its activity is strongly inhibited by pCMB but not by chelating agents.

The third class of neutral aminopeptidases in plants are the proline aminopeptidases or iminopeptidases; these exopeptidases have been noted in crude extracts from a wide variety of dicots and monocots.[27,49,75-79] To date only two higher plant and two algal iminopeptidases have been biochemically purified.[78-82] Proline aminopeptidases have very strong preference of hydrolyzing Pro-β-naphthylamide or Pro-p-nitroanilide but not chromogenic substrates conjugated to any other amino acid. The wheat and *Euglena* iminopeptidases also hydrolyze β-naphthylamide conjugated to hydroxyproline, an amino acid prominent in abundant cell wall proteins of higher plants.[79,81,83] When peptide substrate cleavage was examined in the plant and algal iminopeptidases, significant differences between the higher plant and *Euglena* enzymes were noted. While the apricot and wheat iminopeptidases hydrolyze a wide range of dipeptides with N-terminal Pro, they do not cleave the dipeptide Pro-Pro. In contrast, the only dipeptide cleaved by the *Euglena* iminopeptidase is Pro-Pro; other Pro-X peptides, where X is any other amino acid residue, are not hydrolyzed.[81] Besides the strong substrate bias, it appears these exopeptidases share other properties such as high molecular weight (168-440 kDa) and inhibition by pHMB but not by chelating agents. A 225-kDa iminopeptidase was partially purified from *Chlamydomonas reinhardii*; unlike the *Euglena*, wheat and apricot iminopeptidases, the *C. reinhardii* exopeptidase has an alkaline pH optimum.[82] Genes encoding iminopeptidases have been isolated from several prokaryotes. Protein sequence comparisons revealed conserved regions within these exopeptidases; in addition, based on catalytic site consensus sequences and sensitivity to peptidase inhibitors, at least two classes of prokaryotic iminopeptidases were identified.[84-85] The relationship of the plant iminopeptidases to these prokaryotic enzymes and to the mammalian proline aminopeptidase will only be resolved upon determination of the degree of epitope and sequence conservation in the plant, animal and prokaryotic enzymes.[44,86]

ALKALINE AMINOPEPTIDASES

Alkaline aminopeptidase activities have been detected in crude extracts of a wide variety of dicots including peanuts, soybean, pea, kidney bean, sugarbeet, spinach, squash and cabbage.[50,72,75,87-91] Alkaline aminopeptidases were also detected in a gymnosperm, Scots pine, and in the monocots, *X Haynaldoticum sardoum*, oats, wheat, barley and maize.[59,73,92-95,97] Most alkaline aminopeptidases hydrolyze peptide, β-naphthylamide and p-nitroanilide substrates (Table 8.3). However, several alkaline aminopeptidases, like the kidney bean and barley LAPs and the barley, kidney bean and squash dipeptidases, only hydrolyze peptides efficiently.[87,88,94-96]

A moderate number of alkaline aminopeptidases have been partially or completely purified (Tables 8.3, 8.4). It is clear that the alkaline aminopeptidases include enzymes with a wide range of properties. The seed and leaf leucine aminopeptidases (described below; Table 8.4) are the best characterized alkaline aminopeptidases and form a discrete group within this division. Twelve other alkaline aminopeptidases have been isolated from higher plants or algae and are not analogous to the plant and animal LAPs; they represent a diverse array of activities that, for the most part, are incompletely characterized (Table 8.3). Many alkaline aminopeptidases are sensitive to sulfhydryl reagents like pCMB or pHMB, and some are metalloenzymes.[59,75,87,91,94,96] At the present time, comprehensive substrate specificity studies for the alkaline aminopeptidases are lacking; these studies should include a wide range of amino acyl-β-naphthylamide, -p-nitroanilide and peptide substrates. While many of the alkaline aminopeptidases were shown to be active on peptide substrates, the *Euglena*, oat, *X H. sardoum*, *Chlamydomonas* and pea enzymes must still be considered naphthylamidases.[68,73,82,92,97]

The Mikola and Mikola system for classifying aminopeptidases according to pH optima gives the plant scientific community a solid framework with which to compare less well-characterized aminopeptidases. It also has some limitations. It is clear that both the neutral and alkaline aminopeptidase classes include enzymes with diverse substrate preferences, molecular masses and divalent cation requirements. In some cases, the pH optima for a particular enzyme is dependent on the reaction conditions or substrates tested. For example, the barley LAP has different optimal pHs for hydrolyzing the dipeptide Leu-Tyr in sodium phosphate buffer (pH = 8.5) and sodium borate buffer (pH = 10.5).[95] In addition, some aminopeptidases isolated exhibit a substrate-dependent pH optima.[55,59,61] The barley "Phe" aminopeptidase pH optima for hydrolysis of Phe-β-naphthylamide and peptides are 7.2 and 5.8-6.5, respectively.[61] Finally, strict categorization of exopeptidases according to their pH optima may cause misclassification of exopeptidases. For example, all three *Chlamydomonas* aminopeptidases have slightly alkaline pH optima (Table 8.3).[82] However, when the substrate specificities of these enzymes were examined, all three enzymes were more similar to the neutral aminopeptidases of higher plants (Tables 8.1, 8.2). Ultimately, classification of the neutral and alkaline aminopeptidases must be based upon immunological studies that demonstrate epitope conservation and cell compartment localization, and peptide sequence similarities. Given these criteria, two classes of plant aminopeptidases can unequivocally be identified based on similarities to well-characterized mammalian exopeptidases: the plant leucine aminopeptidases and cathepsin H homologs.

PLANT LEUCINE AMINOPEPTIDASES

One of the most discriminating and informative classification systems for exopeptidases is based on immunological cross reactivity, peptide sequence similarity and conservation of catalytic and ion-binding sites.[39-41] Demonstration of a high amino acid sequence similarity in selected domains or throughout the peptidase may reflect important similarities in catalytic mechanisms, protein structure and perhaps biological functions. Based on these most stringent criteria, one set of plant alkaline aminopeptidases was shown to be members of the bovine lens leucine aminopeptidase family.[34,35,37]

The mammalian leucine aminopeptidase is one of the best characterized aminopeptidases at the physiological, biochemical and structural level (see Chapter 2).[7,98,99] The bovine lens LAP is a hexameric enzyme with MW of 327 kDa and has 6 identical subunits of 54 kDa.[100] In humans, two LAPs are detected in liver. One enzyme is similar to the bovine LAP, since it is a homomultimer with 55 kDa subunits; the second LAP is a heterohexamer with three 53 kDa and three 65 kDa subunits.[46,101] The bovine LAP is a metalloenzyme with each subunit binding two zinc ions.[102] The three-dimensional structure of the bovine LAP and its complex with the inhibitors bestatin and amastatin have recently been determined by x-ray crystallography, and a catalytic mechanism has been proposed based on the binding modes of bestatin.[41,103,104] The Zn-ion binding and substrate-binding sites have also been elucidated.[105]

The amino acid sequence of the bovine LAP is available from direct sequencing of the lens LAP and deduced from the DNA sequence of a kidney *lap* cDNA clone.[106,107] cDNAs encoding a constitutively expressed LAP from *Arabidopsis* leaves and wound-induced LAPs from potato and tomato leaves have been recently isolated and characterized.[34,35,37] The plant LAP protein sequences show a striking degree of similarity (52-56%) with animal and prokaryotic aminopeptidases (Table 8.5). Sequence comparisons indicate that most divergence occurs in the N-terminal domain of LAP. When the C-terminal regions are compared, the plant LAPs are more than 70% similar to the bovine LAP; all residues shown to be important in zinc ion binding and substrate binding in the bovine lens LAP are absolutely conserved.[34] When LAP peptide sequences from the plant kingdom are compared, higher amino acid similarity is observed; for example, the tomato and potato LAPs are 95% similar at the amino acid level. This high degree of conservation is in accordance with their evolutionary relatedness since both tomato and potato belong to the Solanaceae.

Table 8.5. Percentage amino acid similarity between animal, plant and prokaryotic LAPs

	Tomato	Potato	*Arabidopsis*	Bovine lens	*E. coli*	*Rickestia*
Tomato	100	95.6	81.5	53.2	55.1	53.2
Potato		100	82.2	52.4	56.1	55.5
Arabidopsis			100	55.3	56.2	54.8
Bovine lens				100	56.0	55.4
E. coli					100	60.0
Rickestia						100

The first plant leucine aminopeptidase cDNA clone was isolated from *Arabidopsis* after screening an expression library using polyclonal antibodies against purified plasma-membrane proteins.[37] Although the manner in which the *Arabidopsis* clone was isolated suggested that LAP is membrane-associated, subsequent studies indicated that it is likely to be a soluble enzyme.[108] Immunoblot studies showed that the 55 kDa LAP is present in all developmental stages and in all organs examined and is not induced in response to exogenous phytohormones or mechanical wounding.[108] Furthermore, cross reactive LAP proteins are detected in leaves of fourteen different higher plants; LAP proteins are not detected in *Saccharomyces cerevisiae* or in the ascomycetes, *Anthoceros crispulus*.

In contrast to the *Arabidopsis* LAP, the tomato and potato *Lap* cDNAs encode mRNAs that are not found in healthy leaves and are induced by mechanical wounding or exogenous wound signals.[34,35] The potato *Lap* cDNA clone was originally identified as encoding a mRNA that increases in response to the phytohormone abscisic acid which is known to regulate both wound and water-stress responses in plants.[35,109,110] In addition, the potato *Lap* mRNA accumulates in response to jasmonic acid (a regulator of the wound response and senescence) and mechanical wounding. Induction of *Lap* mRNA after addition of exogenous jasmonic acid correlates with an increase of exopeptidase activity in crude extracts.[111] Inspection of the N-terminal sequence of the deduced LAP proprotein reveals that it shares many features commonly found in chloroplast transit peptides, and chloroplast localization has been suggested.[111]

Using a subtractive hybridization strategy to study changes in tomato gene expression following *Pseudomonas syringae* pv. tomato infection, Pautot et al identified a pathogen-induced *Lap* cDNA clone.[34] Expression studies indicated that *Lap* mRNA is induced in response to mechanical wounding, bacterial infection and insect infestation, and *Lap* mRNA levels are strictly correlated with changes in LAP protein levels and a high molecular weight (>327 kDa) aminopeptidase activity.[34,112a] Immunoblot studies, using polyclonal or affinity-purified antibodies to the tomato LAP protein, demonstrated that there is a complex array of LAP-related proteins in tomato leaves;[112a] four classes of cross reactive proteins are detected. A cluster of acidic 55 kDa polypeptides (pI of 5.9) is induced in response to wounding and pathogen infection, whereas the 55 kDa LAPs with a neutral pIs and two larger LAP-like proteins are detected in both healthy and wounded leaves. Although the 66 and 77 kDa LAP-like proteins share antigenic determinants with the wound-induced and constitutive LAPs from tomato leaves, it is not clear if they function as aminopeptidases; a LAP-related protein has also been noted in human cells.[112] The tomato LAP protein with the neutral pI appears to be present in many plant species and most likely corresponds to the LAP identified in *Arabidopsis* (Chao and Walling, unpublished results).[37,108] The acidic LAP proteins appear to correlate with the wound-induced LAP of potato.[35,111]

There are two genes in the tomato genome that encode the wound-induced LAPs.[34] The genes that encode the neutral 55 kDa LAPs of tomato are sufficiently diverged from the wound-induced *Lap* genes, since genomic DNA blots at reduced stringencies do not allow identification of cross-hybridizing bands (Tu and Walling, unpublished results). Transgenic tomato and tobacco plants expressing *Lap* promoter-GUS fusion genes have been constructed, and the analysis of these plants will aid in determining the developmental and environmental signals that are essential for the expression of the wound-induced *Lap* genes (Chao, Pautot and Walling, unpublished).

The strong conservation of the leucine aminopeptidases in prokaryotes, plants and animals suggests that they are derived from an ancient ancestral protein and therefore may exhibit similar biochemical properties. The *Arabidopsis*, potato and tomato LAP proteins have been overexpressed in *E. coli* (Gu and Walling, unpublished results).[37,111] Some of the biochemical properties of these plant LAPs were determined in crude bacterial extracts. Similar to the animal and prokaryotic LAPs, the potato and *Arabidopsis* LAPs have alkaline pH optima and are heat stable.[37,111] These enzymes are inhibited by chelating agents (EDTA or 1,10-phenanthroline) and zinc ions, and are stimulated by manganese and magnesium ions. The potato and tomato LAPs are inhibited by bestatin, and the *Arabidopsis* and tomato LAPs appear to function as multimeric complexes with molecular weights greater than 327 kDa.[108,112a] Analysis of the wound-induced tomato LAP showed that the enzyme is a multimer with 55 kDa subunits.[112a]

Given the diverse strategies used for the isolation of the plant LAP clones and the differential expression of LAPs during plant development and in response to stress, it is likely that the plant constitutive and wound-induced LAPs may display significant differences in substrate utilization. Since enzyme kinetic studies for the *Arabidopsis*, potato and tomato LAPs are lacking, their relative preference for dipeptides, tripeptides, amino acyl-β-naphthylamide or -p-nitroanilide substrates has yet to be determined. Relative to peptide substrates, the bovine and swine LAPs are known to use amino acyl-p-nitroanilide and -β-naphthylamide substrates inefficiently.[7] The tomato, *Arabidopsis* and potato LAPs all hydrolyze Leu-β-naphthylamide and/or Leu-p-nitroanilide;[112a] the tomato LAP also readily hydrolyzes leucinamide.[37,111,112a] Further characterization of the *Arabidopsis* (Bartling, personal communication) and tomato LAPs is ongoing (Gu and Walling, unpublished results). Herbers et al showed that crude extracts from *E. coli* overexpressing the potato LAP hydrolyzed Arg- and Met-p-nitroanilide 4.6- and 1.2-fold more efficiently than Leu-p-nitroanilide.[111] It has yet to be determined if the substrate specificities identified by nitroanilide and naphthylamide substrates reflect plant LAP activities on peptides.

Two aminopeptidases with enzymatic and biochemical properties closely resembling animal LAP have also been identified and purified from germinating barley seeds and resting kidney bean cotyledons.[88,95] Barley LAP was the first plant aminopeptidase activity to be identified.[19] In 1975, Sopanen and Mikola biochemically purified this enzyme from germinating barley seeds, and it became clear that the barley aminopeptidase has many characteristics that are similar to the bovine LAP.[7,95] First, the barley LAP hydrolyzes tripeptides (Leu-Gly-Gly and Met-Leu-Gly) and dipeptides (Leu-Tyr and Leu-Gly) efficiently, whereas leucinamide and Leu-β-naphthylamide are hydrolyzed 6- and 1600-fold less efficiently than Leu-Gly-Gly.[95] Second, the barley LAP is a multimeric enzyme with a molecular weight similar to the bovine LAP, has an alkaline pH optimum (pH 8.5 to 10.5), and is stabilized by magnesium and manganese ions.

A second enzyme with biochemical properties similar to the animal LAP was recently isolated from kidney bean cotyledons.[88] Similar to the animal and barley LAPs, the kidney bean LAP has a pH optimum of 9.0 and efficiently hydrolyzes tripeptides and dipeptides with N-terminal leucine residues. Like the mammalian LAPs, bestatin is a potent inhibitor, and magnesium ions significantly stabilize the enzyme. The kidney bean LAP is a multimeric enzyme with a molecular weight of 360 kDa, but with respect to subunit composition, the kidney bean LAP is most similar to the human LAP characterized by Kohno et al.[46] Unlike the bovine and barley LAPs that are homohexamers, the kidney bean and human LAPs are heterohexamers with equimolar amounts of two distinct-sized subunits. The kidney bean subunits are 58 and 60 kDa in size, while the human LAP has subunits of 53 and 65 kDa. Since cDNA clones for the kidney bean and barley seed LAPs have not been isolated or characterized, evolutionary relationships between the seed LAPs and the *Arabidopsis*, tomato and potato leaf LAPs are not possible at the present time. In addition, the relatedness of the subunits from the heteromultimeric LAPs and the 55 kDa subunits from the homohexameric LAPs needs to be determined. The isolation of antibodies to the kidney bean LAP will aid in the elucidation of these relationships (Mikkonen, personal communication).

PLANT CATHEPSIN H HOMOLOGS

The second class of aminopeptidases that have been cloned are the peptidases that show similarity to the lysosomal cathepsin H. Two cathepsin H homologs have been identified in higher plants. Aleurain shares 65% amino acid sequence identity with the proteolytic domain of the human cathepsin H.[36] Aleurain is a thiol endoprotease that is induced after germination in barley seeds, but is present in all organs of the plant. Aleurain is synthesized as a 44 kDa proprotein that is processed to a mature 32 kDa peptidase. Immunolocalization studies have shown that aleurain is localized to the vacuole (the analog of the animal lysosome) in the seed aleurone cell layer.[113] Recently, Holwerda and Rogers showed that aleurain, like cathepsin H, hydrolyzes aminopeptidase substrates more effectively than endopeptidase substrates.[114,115] While limited substrate specificity data is available for this aminopeptidase, it is clear that it can efficiently hydrolyze substrates with N-terminal Arg, Phe and Leu residues.[114,116] Given aleurain's vacuolar localization, it is not surprising that the aminopeptidase is stable at acidic pHs. Somewhat surprising is the fact that optimal aminopeptidase activity actually occurs in the neutral pH range.[114] Recently, a cathepsin H homolog (oryzain γ) was cloned from rice; the deduced amino acid sequence had 85% and 60% similarity with barley aleurain and cathepsin H, respectively.[117] Its activity as an aminopeptidase has yet to be demonstrated.

ROLE OF PLANT AMINOPEPTIDASES IN PLANT DEVELOPMENT

Aminopeptidase activities are found in all organs of higher plants. Most plant aminopeptidases are found in many diverse organs, and their levels vary quantitatively.[59,108] There are only a few examples of organ-specific aminopeptidases or aminopeptidases induced in response to environmental cues.[28,34,35] While the roles of single aminopeptidases have not yet been elucidated, there is a strong body of plant literature correlating the presence of aminopeptidase activities with stages in growth and development. In some studies, multiple aminopeptidase activities were followed using in situ gel assays or by assaying partially purified enzymes; this allowed some limited insights into the coordinate expression of the different classes of aminopeptidases and their probable role in modulating cell metabolism and protein turnover. Most often, a single aminopeptidase activity in crude tissue homogenates was monitored; these experiments have yielded limited data since more than one aminopeptidase may hydrolyze the substrate(s) used in the assay. But given the wide number of species in which similar studies were conducted, several unifying themes about the role of aminopeptidases in plant growth and development have emerged.

SEED GERMINATION

The search for aminopeptidase activity in seeds was initially driven by the importance of the mobilization of storage proteins for malt production in beer production. Therefore, aminopeptidases were identified in barley seeds as early as 1929.[19] With the identification of endoproteases in germinating seeds and the realization that some plant exopeptidases had properties similar to those found in a diverse array of animals, the quest to elucidate the contribution of aminopeptidases, carboxypeptidases and endoproteinases in the hydrolysis of seed storage proteins continued.[7,19,22] The mobilization of seed protein reserves from storage sites provides a major source of carbon, nitrogen and sulfur for the developing seedling. After seed germination, an increase of proteolytic activity is strongly correlated with a decrease of seed storage proteins; storage proteins are hydrolyzed to their constituent amino acids or to dipeptides and are subsequently transported and absorbed by the actively growing shoot and roots.[21,22,118,119] The change of proteolytic activities after seed germination has been widely studied in both monocots and dicots and to a more limited extent in gymnosperms.[21,22,73,120] Depending on the plant species and the properties of the storage proteins, the contribution of endopeptidases, aminopeptidases and carboxypeptidases to the degradation of storage proteins differs.

High levels of aminopeptidases are found in resting mature seed. However after germination, the changes in aminopeptidase activity in the storage organs varies tremendously; in some cases, cultivar-dependent differences were noted within a species, thereby making generalizations difficult.[121] In several plant species, aminopeptidase activities increase after germination and strongly correlated with the degradation of protein reserves. This correlation has been made for protein mobilization from cotyledons of lettuce and jojoba, the endosperm of castor bean, the megagametophyte of gymnosperms and from fern spores.[73,77,120,122-124] In most cases, the biochemical nature of these aminopeptidases is not well characterized.[77,122,124] In three cases, multiple aminopeptidase activities were detected in the storage organs. In castor bean, leucine naphthylamidase and iminopeptidase activities rise coordinately after germination, thereby implicating both activities in the hydrolysis of endosperm storage proteins.[71] In contrast, differential regulation of aminopeptidases was documented in soybeans and Scots pine.[54,73] Four aminopeptidases are present in germinating Scots pine seeds.[73,125] The two alkaline aminopeptidases that hydrolyze the dipeptides Leu-Tyr or Ala-Gly increase dramatically after germination, and their activities decline only when greater than 90% of the storage reserves are depleted. The two naphthylamidase activities of Scots pine are maintained at the same level present in resting seed and decline at the same time as the Leu-Tyr and Ala-Gly aminopeptidases. In soybean, two aminopeptidase activities are detected. One aminopeptidase declines and a novel aminopeptidase increases after imbibition.[54] It is unclear if one or both of the soybean aminopeptidases have important roles in seed protein mobilization.

Although the studies described above suggest a role for aminopeptidases in seed protein mobilization, studies with other monocots and dicots suggest that this may not be a universal role for aminopeptidases in plants. For example, in many plants aminopeptidase activity levels decrease prior to the digestion of the majority of the storage protein reserves. For example, Chrispeels and Boulter found that a leucine naphthylamidase activity declines markedly in the cotyledons of mungbean upon imbibition.[126] Similar decreases in aminopeptidase activity after germination have also been observed in other dicots including pea and kidney beans.[52,127,128] The lack of correlation with aminopeptidase activity and seed protein degradation was also noted in three monocots: barley, wheat and maize. Aminopeptidase activities in crude extracts of maize and wheat seed decline after germination and preceded storage protein mobilization from the endosperm.[28,129] When germinating barley seed was examined five aminopeptidases and three carboxypeptidases were detected.[130] Three of the aminopeptidases are naphthylamidases with neutral pH optima that prefer substrates with Phe, Leu or Arg residues. The remaining aminopeptidases have alkaline pH optima and are a leucine aminopeptidase and an Ala-Gly dipeptidase. Cell fractionation studies indicate that the five aminopeptidases are detected in the two storage protein repositories of the barley grain (the aleurone and endosperm), and the alkaline exopeptidases are soluble enzymes.[130] Similar to peanut (a dicot), there is no change in the level of the five aminopeptidase activities in the barley endosperm during the 5 days after imbibition; during this time carboxypeptidases increase approximately 5-fold.[72,130] Since there is no rise in the five aminopeptidase activities in the barley endosperm and the three naphthylamidases are present at higher levels in the embryo than in the endosperm, it was assumed that these enzymes had a negligible role in mobilization of storage reserves from the endosperm.[130] Alternatively, it is possible that the high levels of the five aminopeptidases that are present in the resting seed are sufficient to facilitate the degradation of the endosperm storage proteins, and further induction is not necessary.

The dipeptides or single amino acid residues that result from hydrolysis of proteins in the endosperm or aleurone are transported to and absorbed by the scutellum, the monocot organ that supplies nutrients to the seedling's shoot and root systems.[118,119] High levels of aminopeptidases are detected in the scutellum after germination; these activities are likely to degrade endogenous scutellar proteins and complete hydrolysis of the endosperm and aleurone-derived peptides for utilization by the rapidly dividing cells in the developing shoot and roots.[130,131]

Since in most plants, aminopeptidase gene probes and antibodies are lacking, it is not clear if the changes in seed aminopeptidase activities after imbibition are controlled by alterations in the rates of gene transcription, mRNA processing or turnover, protein synthesis or turnover, post-translational modifications, or changes in the level of aminopeptidase inhibitors or activators. To date, there are only two plants where these regulatory processes in the seed have been investigated. First, immunoblot studies showed that the decline in mungbean aminopeptidase activity after germination is well correlated with the decline in protein levels, thereby eliminating translational or post-translational events as key regulatory steps.[58] Second, changes in aleurain mRNAs are strictly correlated with the changes in the levels of aleurain protein and activity; after germination these changes are induced by the phytohormone gibberellic acid.[86,132] Several other investigators have examined the importance of shoot-derived signals (most probably phytohormones) in the changes in seed aminopeptidase activities, and species-specific differences were noted. Removal of the shoot axis does not alter the aminopeptidase levels in chick-pea cotyledons.[133] In contrast, the developing shoot of lettuce and mungbean seedlings supplies an important signal for the germinating seeds, for the removal of the shoot axis blocks the rise in aminopeptidase activity in cotyledons that normally follows germination.[122,134] The "signal" from the lettuce shoot axis is light-dependent, since aminopeptidase levels do not rise in seeds germinated in the dark. It is clear that there are many questions to be addressed regarding the mechanisms that regulate seed aminopeptidase activities.

SEED AND FRUIT MATURATION

While there are a large number of studies examining the changes in aminopeptidases in storage tissues after germination, only four studies have documented changes in aminopeptidase activities during seed and fruit maturation. In pea, maize, wheat and *X H. sardoum*, a rise in aminopeptidase activity occurs early in seed maturation, and activities decline with the onset of seed desiccation.[28,29,52,121] The high level of aminopeptidase activities during early phases of seed maturation is correlated with a period of rapid cell division and differentiation. It is possible that these metabolically active cells require higher levels of aminopeptidases to ensure efficient turnover of damaged or inactive proteins. Alternatively, aminopeptidases may be involved in the final hydrolysis of peptides to ensure adequate amino acid pools for the synthesis of the abundant storage proteins that occurs in the middle and late stages of seed maturation.

The changes in aminopeptidases during seed ripening is best characterized in maize where four aminopeptidases (AMP1-4) reach peak levels in the developing endosperm during the period of rapid cell divisions.[28,59] Each of the four enzymes have different substrate specificities. AMP1 and AMP4 are neutral aminopeptidases that preferentially hydrolyze substrates with basic and aromatic amino acids, respectively. AMP2 and AMP3 have alkaline pH optima and hydrolyze Ala/Gly or basic residues, respectively, from β-naphthylamide substrates. AMP2 is an endosperm-specific aminopeptidase, while AMP1, AMP3 and AMP4 are found in the endosperm, aleurone and scutellum in maturing seeds and in all organs of the mature maize plant. To date AMP2 is the only organ-specific aminopeptidase identified in higher plants, and its specific role in the metabolism of immature endosperm has yet to be elucidated.

The protective layers of cells that surround the developing seed also have aminopeptidases. For example, aminopeptidases are detected in the seedcoats of pea and siliques of *Arabidopsis*.[108,135] In fleshy fruits, like peach, the mesocarp surrounds the seed during ripening. In early stages of fruit development, the mesocarp proceeds through a period of rapid cell divisions which is followed by a period of ripening characterized by the accelerated turnover of proteins and synthesis of new proteins that facilitate the changes in texture, skin and flesh color, softening, flavor and sweetness.[136] In peach, a leucine arylamidase is present at high levels in the mesocarp during the period of rapid cell divisions and the period of "softening" and protein turnover.[137]

SEEDLING GROWTH AND DEVELOPMENT

Aminopeptidases have also been monitored during seedling growth and development. While the roles for aminopeptidases in seedling development cannot be rigorously addressed until mutants that alter aminopeptidase levels in the seedling are isolated and characterized, there are many correlations that are important to note. Many of the aminopeptidases detected after germination or during seed maturation are also found in organs of the developing and mature plant. The ubiquitous expression of these exopeptidases suggests they are important for the protein turnover that is essential for cell maintenance. While most aminopeptidases are widespread, there are significant differences in the levels of each aminopeptidase in specific organs; these quantitative differences may reflect more specific functions.

The changes in aminopeptidase activities in the developing seedling have been most extensively studied in pea, maize, oats and wheat.[27,28,51,129] Generally, aminopeptidase levels are highest in young actively growing tissues; peak activities are generally correlated with the peak of total protein accumulation and fresh weight in an organ.[28,129] In the early stages of shoot and root development that occur immediately after germination, aminopeptidase and endopeptidases activities rise. It is in these cells that rapid cell divisions and differentiation are occurring. As the seedling leaves and roots age and the proportion of cells in these organs that are actively undergoing divisions declines, aminopeptidase activities decrease, while carboxypeptidase activities actually increased.[131] When aminopeptidases are monitored in mature roots, the highest levels are detected in the root tips where cell divisions occur continuously.[130,131,138]

Studies in pea, maize, wheat and oats have demonstrated that multiple aminopeptidase activities are detected in the seedling's leaves and roots, and each aminopeptidase activity appears to be independently regulated.[27,28,51,129] For example, Sopanen and Lauriere monitored aminopeptidase, endoprotease and carboxypeptidase changes in the first leaf of wheat for 19 days (prior to the onset of senescence).[93] During this period four different aminopeptidases were monitored (an Arg and a "Phe" naphthylamidase, as well as Leu-Tyr and Ala-Gly aminopeptidases). While the Arg naphthylamidase activity does not change during the 19 d period, the "Phe" and Ala-Gly aminopeptidase activities increase slightly, and the Leu-Tyr activity increases dramatically. Although it is clear that aminopeptidases are at peak levels in the most metabolically active cells, their specific roles are not understood. These aminopeptidases may be important in ensuring that proteins are recycled efficiently to provide adequate amino acid pools for protein synthesis in these active cells. Alternatively, increased aminopeptidase levels may be essential for the requisite turnover of regulatory proteins that are critical determinants for tissue and organ differentiation and maintenance.

In addition to the events of cell differentiation and cell expansion that occur during seedling development, proteases, including aminopeptidases, may be involved in the differentiation of plastids. Depending on the organ and the developmental and environmental signals present, proplastids differentiate into distinct forms such as chloroplasts, chromoplasts and amyloplasts.[139] Each form has different ultrastructure, lipid and protein composition, and function; furthermore, the distinct plastid forms are interconvertible under different developmental and environmental conditions. For example, upon germination in the dark, proplastids develop into etioplasts that contain a crystalline-like structure called the prolamellar body. Upon exposure to light, the etioplast and its prolamellar body redifferentiate, and a photosynthetically competent chloroplast with its extensive thylakoid membrane network is formed. In this transition, the prolamellar proteins must be turned over, and chloroplast proteins must be synthesized. The turnover of the plastid proteins, in this plastid transformation and others, is presumed to be due to the concerted action of endoproteases and aminopeptidases, since carboxypeptidases have yet to be detected in plastids.[140]

Proof for the localization of aminopeptidases in chloroplasts appeared in 1982 after Waters and colleagues performed a rigorous study of wheat leaves to identify and characterize plastid aminopeptidases.[140] Because of the rigid cell wall and large volume occupied by the central vacuole, purification of cell organelles and cytoplasm without contamination from the vacuole components is difficult. The use of protoplasts to isolate physically intact cell organelles, including vacuoles, aided in determining aminopeptidase localization in plant cells. Using protoplasts and a number of organelle-specific markers in cell fractionation studies, Waters et al unequivocally identified three aminopeptidases in wheat chloroplasts and cytosol; these studies also indicated that there is no detectable aminopeptidase activity in peroxisomes, mitochondria or the central vacuole of wheat leaves.

Similar studies were performed in pea to identify three aminopeptidases in the stromal compartment of the chloroplast.[75] The three pea exopeptidases have a slightly alkaline pH optima (pH = 7.7) and substrate specificity studies using 11 different β-naphthylamide substrates indicate the aminopeptidases correspond to three of the neutral aminopeptidase classes: the proline, "Phe" and "Ala-Leu" aminopeptidases (Tables 8.1, 8.2).[75] Chloroplast-localized aminopeptidases have also been identified in sugarbeet cotyledons and in barley and oat leaves.[23,89,92] Finally, the chloroplast-localization of the wound-induced LAP of potato and tomato has also been suggested, since the N-terminus of the potato LAP has features indicating it may serve as a chloroplast transit peptide.[111,141a] Its alkaline pH optimum is also consistent with the plastid localization. However, it should be noted that the mammalian LAP has an N-terminal propeptide and has an alkaline pH optimum and is a cytosolic protein. Preliminary cell fractionation studies in tomato indicates that the tomato wound-induced LAP is a soluble enzyme, and only a small fraction of the wound-induced LAP protein is localized to the plastid.[112a] The localization of the wound-induced LAPs of potato and tomato must await definitive immunogold localization studies.

While plastid aminopeptidase activities have been identified, their exact role in plastid biogenesis and organelle maintenance is not completely clear. Not only may these exopeptidases be involved in facilitating the interconvertibility of plastid forms during the plant life cycle, aminopeptidases may play important roles in the maturation of chloroplast proteins and the turnover of oxidatively damaged proteins. It is well established that proteolysis is an essential component for the import of nuclear-genome encoded polypeptides that are targeted to the plastid. While there is an intensive effort focused on identifying the enzymes essential for transit peptide processing in plastids, virtually nothing is known about the exopeptidases that are essential for chloroplast protein maturation.[141] It is clear from the comparison of the mature N-termini of a number of chloroplast-localized proteins and the deduced peptide sequences from genomic or cDNA clones that aminopeptidases are active in plastids. In most cases, the N-terminal methionine is removed from chloroplast-encoded proteins including the *psbH* 10 kDa phosphoprotein, cytochrome b-559, the epsilon subunit of the ATP synthase, and the D1 and D2 proteins of the photosystem II reaction center.[142-145] The N-terminus of the spinach CPa-3 polypeptide was processed more extensively, for 14 N-terminal residues were removed to give rise to the mature CPa-3 polypeptide.[142] The influence of aminopeptidase processing on the activity of these proteins and their in vivo half-lives has yet to be determined.

Given the high levels of endoproteases and aminopeptidases in chloroplasts, the aminopeptidases are likely to act in concert with the endoproteases to facilitate the turnover of oxidatively damaged proteins (like D1) or other proteins known to have extremely short half-lives in the light (i.e., NADPH-protochlorophyllide oxidoreductase and phytochrome).[139] In this sense, the plastid and cytosolic aminopeptidases may function in a manner analogous to the lens aminopeptidases of mammals.

Aminopeptidases are also proposed to participate in the plant's response to prolonged growth in the dark; this response is triggered primarily due to carbohydrate starvation and results in an irreversible block in the etioplast to chloroplast transition and a rise in plastid aminopeptidase activities.[89] The rise in cotyledonary aminopeptidase activities in carbohydrate-starved seedlings may be reflective of the salvaging of proteins from etioplasts that are blocked from their normal differentiation program. In contrast, glucose starvation caused a decline in the aminopeptidase activities in maize root tips.[146]

Finally, aminopeptidases may have a role in salvaging nutrients from senescing leaves. Except for leaf senescence induced by environmental stress, natural leaf senescence is a process that is tightly regulated by developmental signals when a plant shifts from vegetative to reproductive growth; for this reason, natural leaf senescence should be considered an important phase of leaf development.[147] Leaf senescence alters many metabolic pathways. Changes in the number of chloroplasts per cell, chloroplast ultrastructure and the significant increase in protein degradation are hallmarks of senescence.[148] Since each cell harbors approximately 100 chloroplasts and it is estimated that chloroplast-localized proteins can account for up to 50% of the leaf proteins, a large proportion of the proteins being degraded must be chloroplast localized. The degradation of leaf proteins during senescence is thought to facilitate the remobilization of the nitrogen, carbon and sulfur reserves from leaves and translocation of these nutrients into the reproductive structures such as seeds and fruits.

While the rise in endoprotease activity is well documented, the role for aminopeptidases in leaf senescence has been harder to evaluate, for the changes in aminopeptidase activities during senescence in higher plants are variable. For example, there is a good correlation with a rise in three aminopeptidase activities in *Euonymous* leaves during the period of leaf senescence (late summer and early fall).[53] In contrast, studies in barley leaves showed that one cytosolic naphthylamidase and three chloroplast activities do not change substantially during leaf senescence; in fact, the activities of the four aminopeptidases are at lower levels in senescing than in young leaves.[23] It is hard to evaluate if this apparent lack of correlation is reflective of the fact that in some plants aminopeptidases do not contribute to the catabolism of proteins during senescence. Alternatively, it is possible that the low levels of the aminopeptidases are sufficient to facilitate protein turnover in senescing leaves or that a class of aminopeptidases (with different substrate specificities), that rise and are essential for protein turnover during senescence, have yet to be identified.

While natural leaf senescence is known to be a genetically programmed process, very little is known about the signals that stimulate the increases in proteolytic enzymes and other changes in gene expression.[149] There are no studies investigating the impact of jasmonic acid, polyamines, ethylene and other plant growth regulators on aminopeptidase expression during senescence. While the tomato wound-induced LAPs are induced by exogenous application of jasmonic acid in young leaves (Chao and Walling, unpublished results), *lap* mRNAs and LAP enzymatic activity are not detectable during leaf senescence (Pautot, personal communication).

FLOWER DEVELOPMENT

There is surprisingly little data available about the expression of aminopeptidases during floral development. Monomeric aminopeptidase activities (AMP1, AMP3 and AMP4) are detected in the anthers, pollen and the silks of maize, and the hexameric *Arabidopsis* LAP is detected in flowers.[28,108] At present nothing is known about the cellular compartment or specific cell types expressing these exoproteases. Since the three maize aminopeptidases and the *Arabidopsis* LAP are also found in all other organs examined, it is likely that these exoproteases are important in the general events of cellular protein turnover. In addition, a rise in aminopeptidase activity was noted during tulip petal senescence, and a rise in the specific activity of an aminopeptidase was noted during pea ovary senescence.[24,25]

A contrasting situation exists for the wound-induced LAPs of potato and tomato. The potato and tomato *Lap* mRNAs are not constitutively expressed; these mRNAs and proteins are only detected in leaves after wounding and exogenous application of wound signal molecules, or during flower development (Pautot, Chao, Holzer and Walling, unpublished results).[35,111] Although the function of wound-induced LAP in flower development is not understood, its expression in the female reproductive structures is consistent with the regulation of a number of other defense-related proteins.[150] It is possible that defense-related proteins accumulate in the developing bud to provide an enhanced defense in the flower that may aid in protecting the eggs and developing embryos from irreversible damage caused by pathogen infection.

ROLE OF AMINOPEPTIDASES IN HOST-PATHOGEN INTERACTIONS

The involvement of peptidases in human disease is well established. Changes in aminopeptidase activities have been correlated with a wide variety of clinical conditions and pathologies in humans.[44] The importance of aminopeptidases in virus recognition and subsequent defense responses was established when it was recognized that the membrane-associated aminopeptidase N serves as a coronavirus receptor, and LAP is induced in cultured human cells by treatment with interferon-γ.[14,15,112]

To curtail pathogen invasion or insect infestation, plants respond in a manner distinct from animals. Plants synthesize a wide array of defense-related proteins that are involved in strengthening the plant cell wall or actively antagonizing pathogens or pests.[151] In addition, in some plant-pathogen interactions, a hypersensitive response is induced. This response occurs in plants carrying a disease-resistance gene when they encounter a pathogen expressing the cognate avirulent gene product. In this interaction, the pathogen synthesizes an elicitor that facilitates rapid recognition of the pathogen by the plant; this triggers an accelerated cell death at sites of infection. Many of the genes induced during the hypersensitive response or in other types of host-pathogen interactions are also induced by mechanical wounding; this reflects the cell damage that is incurred with insect feeding or pathogen ingression into plant tissues. The importance of plant-encoded peptidases in plant-pathogen interactions has only recently been recognized.[34,35,152,153,154,155] The induction of endopeptidases and aminopeptidases in response to wounding, bacterial pathogen or viroid infection, and insect infestation is documented. The action of endoproteases, carboxypeptidases and aminopeptidases in the processing of the wound peptide systemin has also been implicated.[32,33] Finally, treatment of tomato seedlings with the aminopeptidase inhibitor bestatin induces the expression of wound-response genes like leucine aminopeptidase and proteinase inhibitors;[155a] at the present time the identity of the bestatin-sensitive regulator of the tomato defense response is not known.

Increases in aminopeptidases have been noted in response to bacterial pathogens, chewing insects and viroids. Potato spindle tuber viroid-infected leaves contained 3-fold more aminopeptidase activity than healthy tomato leaves.[155] This aminopeptidase is approximately 45 kDa in size, has a neutral pH optimum and is active on Leu- and Phe-p-nitroanilide substrates. This exopeptidase is distinct from the multimeric LAP of tomato that is wound, insect and bacterial pathogen induced.[34] Similar to the tomato LAP, the potato LAP has been shown to be induced in response to mechanical wounding and exogenous application of the wound signal jasmonic acid (Chao and Walling, unpublished results).[35] Unlike the LAPs of tomato, potato *Lap* mRNAs are not induced in response to the bacterial pathogen, *Xanthomonas campestris* pv. *vesicatoria*, and the fungal pathogen, *Phytophtora infestans*.[111] Differences in the regulation of the LAPs of potato and tomato were also noted when the temporal and spatial expression of LAPs after wounding was investigated. While both the potato and tomato *Lap* mRNAs are induced at the site of mechanical wounding, only the tomato *Lap* mRNAs are induced systemically (in upper nonwounded leaves). Since *Lap* mRNAs and proteins accumulate locally and systemically after wounding, LAP could have roles in both regional and systemic defenses which appear to be regulated differentially.[34,112a,156] It is clear that the expression of some defense-related genes that are systemically induced is strictly correlated with the appearance of a nonspecific acquired resistance that protects plants from a broad range of pathogens.[157]

At the present time, it is not clear if the wound-induced LAPs of tomato and potato and the viroid-induced aminopeptidase of tomato serve distinct or overlapping roles in the plant defense response. Since the substrate specificities of the 43 kDa aminopeptidase and the potato and tomato LAPs are distinct, they are likely to act on different peptide and/or protein substrates. Furthermore since subcellular localization of these aminopeptidases is not resolved, their location with the cell may substantially influence the availability of prospective substrates. The degree of aminopeptidase sequestration in organelles may change substantially during plant-pathogen and plant-insect interactions, since regional cell damage and death would release peptidases from all cellular compartments. Under these circumstances, aminopeptidases may directly encounter the pathogen or pathogen-secreted proteins and peptides.

Plant aminopeptidases have been implicated in the processing of pathogen-synthesized peptides. The role of an aminopeptidase in the activation of a bacterial phytotoxin has been suggested. Phaseolotoxin, produced by *Pseudomonas syringae* pv. *phaseolicola*, is a tripeptide with the amino acids N*-N'-sulpho-diamino phosphinyl-L-ornithine (SOrn), alanine and homoarginine. This toxin causes halo blight disease by inhibiting ornithine carbamoyl transferase (OCT) and thereby inducing ornithine accumulation, stunting and chlorosis.[158] Phaseolotoxin is processed in planta to release the modified ornithine residue (pSOrn); while both compounds inhibit the OCT, pSOrn is the most potent irreversible inhibitor.[159] The processing of phaseolotoxin can be mimicked in vitro by the processing of the peptide toxin by the mammalian LAP.[160] This suggests that the constitutive or wound-induced LAP or another aminopeptidase with a similar substrate specificity has an important role in the development of halo blight disease.

A second potential pathogen-derived substrate for aminopeptidases is a race-specific elicitor of *Cladosporium fulvum*. This elicitor is encoded by the *C. fulvum avr9* gene and induces a hypersensitive response in the tomato plants carrying the complementary resistance gene *Cf9*. Avr9 is exported to the extracellular space as a 34 amino acid peptide that is processed in planta to its active form of 28 residues.[161] Since the processing events occur at the N-terminus of the 34-aa precursor peptide, it is likely that a tomato aminopeptidase is involved in the processing events that occur either before or during the recognition of the *avr9* elicitor by the tomato *Cf9* gene product. The ability of the wound-induced LAP of tomato to process the Avr9 peptide is currently being tested (Walling, deWit and Gu, unpublished results).

It is also possible that the defense-related aminopeptidases degrade pathogen proteins or peptides that are essential for pathogen growth and multiplication in plant tissue. For example, the chitinases and β-1,3-glucanases which are induced in the plant defense response actively hydrolyze fungal and pathogen cell walls and thereby deter pathogen growth.[151] It is possible that aminopeptidases may act on pathogen surface proteins to interfere with viability. In addition, in conjunction with proteinase inhibitors I and II, or perhaps independently, aminopeptidases may be involved in the inactivation of proteins or peptide hormones in the insect gut, thereby interfering with insect growth and development.

Plant aminopeptidases may also activate and inactivate proteins or peptides that are important for the regulation (activation or inactivation) of the plant defense response. In this context, plant aminopeptidases would have regulatory roles similar to mammalian aminopeptidases that generate or inactivate peptide hormones. While it is clear that the bioactive wound-peptide systemin is processed at both its N-terminal and COOH-terminal ends, it has yet to be determined if the processing events are facilitated by exopeptidases that are constitutively expressed or are pathogen/wound-induced.[32,33] The role of aminopeptidases in the activation or inactivation of the plant-defense response can be elucidated by determining the substrate specificity of the plant aminopeptidases and evaluation of transgenic plants that overexpress aminopeptidase antisense mRNAs. These experiments are ongoing in tomato (Pautot, Gu, Holzer and Walling, unpublished data). Data from the analysis of transgenic potatoes suggest that the potato LAP does not regulate the defense-response. Given the differences in the expression of the potato and tomato LAPs, the roles for LAP may differ in the two species.[34]

The roles for aminopeptidases discussed above propose the interaction of exopeptidases with specific substrates; in addition, a number of different roles can be proposed that do not require interaction with specific proteins. Aminopeptidases may facilitate accelerated protein turnover in cells responding to pathogen invasion or wounding; in this context, three potential roles of increased aminopeptidase activity can be envisioned. First, aminopeptidases may be involved in the rapid turnover of proteins to increase the amino acid pools for the synthesis of abundant defense-related proteins that accompanies wounding or pathogen ingression. Second, since most cells that act at the site of pathogen invasion respond by localized cell death, it is possible that aminopeptidases act to salvage carbon and nitrogen from cells committed to cell death. Finally, since a transient burst of active oxygen species such as superoxide anion, hydrogen peroxide, and hydroxyl radical accompanies the plant defense response and proteins are easily damaged by these agents, aminopeptidases may rapidly turnover these damaged proteins to facilitate the cell's recovery from oxidative stress.[162,163]

To investigate the role of aminopeptidases in the plant defense response, mutants that eliminate or down-regulate aminopeptidase gene expression must be evaluated. The biological function of the tomato and potato LAPs is being evaluated by the analysis of transgenic plants expressing an antisense *Lap* gene construct. Analysis of the transgenic tomato antisense plants is ongoing (Pautot and Walling, unpublished results). Evaluation of the transgenic potato plants overexpressing an antisense *Lap* mRNA indicated that *Lap* mRNA and LAP protein were not detectable after treatment with the wound signal jasmonic acid.[111] The transgenic potato plants were morphologically normal, and down-regulation of LAP did not influence the induction of other defense-response genes.

PLANT AMINOPEPTIDASES IN RESPONSE TO ABIOTIC STRESS

The evidence for a role of aminopeptidases in plant responses to environmental stress is primarily correlative, but data have accumulated suggesting that further investigation into the roles of aminopeptidases in response to drought, salt-tolerance, cold-stress and ozone should be pursued. In response to high levels of ozone in the environment, active oxygen species accumulate and cause significant damage to plants. It is clear that *Lap* mRNAs are induced in response to ozone in parsley.[164] Their role is likely to facilitate turnover of oxidatively damaged proteins.

The response to high salt, drought and cold involves the plant adjusting to a water deficit. Many of the physiological and biochemical changes are similar in these abiotic stresses, therefore it is not surprising that some genes that are induced in response to drought are also induced by high salt or cold.[109] Changes in amino acid pools have been documented during these stresses and may result from changes in the rates of amino acid biosynthesis, amino acid utilization and/or protein turnover. Given the role of aminopeptidases in releasing single amino acids from proteins and peptides and its ultimate impact on amino acid pools, it is possible that aminopeptidases have an important role in osmotic adjustment in plants. Alternatively, there is evidence for a role of oxidative stress during drought and chilling; therefore aminopeptidases may act to turnover damaged proteins.[165]

To date, there are two studies in rice that showed increased aminopeptidase activities in roots and leaves in response to a saline environment.[166,167] When aminopeptidase activities were compared in salt-sensitive and -tolerant cultivars of rice, tolerant cultivars had increased levels of aminopeptidase activity and increased free amino acids pools. Amino acids may function as compatible cytoplasmic solutes, and increases in their level may facilitate adjustment of the osmotic potential of the cytoplasm during salinity or drought stress.

FUTURE DIRECTIONS

Plants harbor a diverse array of aminopeptidases. While most are detected throughout the plant, there are a small number of aminopeptidases that are organ-specific or are induced in response to environmental stresses which may have more selective roles in protein or peptide metabolism. Most of the data available are correlative, and there are limited data rigorously testing the role of an aminopeptidase in plant development and responses to abiotic and biotic stresses. The impact of each aminopeptidase in the plant life cycle can only be determined upon the characterization of natural or genetically engineered mutants that eliminate or down-regulate aminopeptidase production. A few null mutants for a set of seed aminopeptidases were identified in soybean; these plants were viable, suggesting that either some aminopeptidases in germinating seeds were not essential for plant viability or that there is functional redundancy in the seed aminopeptidase pool that can accommodate the loss of one exoprotease. Future experimentation awaits the isolation of aminopeptidase genes encoding the neutral and alkaline aminopeptidases (Tables 8.1-8.3). With aminopeptidase gene probes in hand and given the extraordinary success of antisense gene technology in higher plants, the exact role of individual aminopeptidase gene products will be elucidated. Since only leucine aminopeptidase and aleurain gene probes have been isolated, this field of experimentation is clearly in its infancy. While antisense technology down-regulates an entire gene family, it does not allow the contribution of specific genes within that family to be evaluated. Recently, a transgenic tomato plant was identified that had one of the *Lap* genes inactivated by the insertion of a transposon (M. van Haren, personal communication). This plant is currently under investigation for its impact on development and host-pathogen interactions; given a greater than 98% identity of the tomato *LapA1* and *LapA2* gene products[112a] it may be unlikely for a discernible phenotype to be detected. Finally, overexpression of aminopeptidases using strong "constitutive" promoters, like the 35S promoter of cauliflower mosaic virus may also yield insights into the roles of aminopeptidases in plant development and responses to biotic and abiotic stresses.

The second avenue of research that is needed is the development of additional plant aminopeptidase gene probes. The deduced amino acid sequences of these genes may allow the identification of animal or prokaryotic analogs and will aid in the development of a strictly defined aminopeptidase classification system. In addition, the production of polyclonal or monoclonal antibodies that recognize plant aminopeptidases will not only aid in the development of a refined classification system but they can be used in immunogold localization studies to unequivocally correlate peptidases and subcellular compartments. While cell fractionation studies have indicated aminopeptidases localized in the cytosol and the plastid; membrane-bound, mitochondrial-localized and extracellular aminopeptidases have yet to be identified in plants.

REFERENCES

1. Miller CR, Kukral AM, Miller JL et al. *PepM* is essential gene in *Salmonella typhimurium*. J Bacteriol 1989; 171:5215-7.
2. Chang Y-H, Teichert U, Smith JA. Molecular cloning, deletion, and overexpression of a methionine AP gene from *Saccharomyces cerevisiae*. J Biol Chem 1992;267:8007-11.

3. Mckelvy JF, Blumberg S. Inactivation and metabolism of neuropeptides. Ann Rev Neurosci 1986; 9:415-34.

4. Kenny AJ, Stephenson SL, Turner AJ. Cell surface peptidases. In: Kenny AJ, Turner AJ, eds. Mammalian Ectoenzymes. New York: Elsevier, 1987:169-210.

5. Steele MK, Myers LS, Deschepper CF et al. Influence of ovarian hormones on the regulation of lutenizing hormone and prolactin release by angiotensin II. In: Johnston CA, Barnes CD, eds. Brain-gut Peptides and Reproductive Function. Boca Raton, Florida: CRC Press, 1991:1-19.

6. Squire CR, Talebian M, Menon JG et al. Leucine aminopeptidase-like activity in *Aplysia* hemolymph rapidly degrades biologically active bag cell peptide fragments. J Biol Chem 1991; 266:22355-63.

7. Hanson H, Frohne M. Crystalline leucine aminopeptidase from lens. Meth Enzymol 1976; 45:504-10.

8. Bachmair AD, Finley D, Varshavsky A. In vivo half-life of a protein is a function of its amino terminal residue. Science 1986; 234:179-86.

9. Varshavsky A. The N-end rule. Cell 1992; 69:725-35.

10. Vierstra RD. Protein degradation in plants. Annu Rev Plant Physiol Plant Mol Biol 1993; 44:385-410.

11. Yoshimoto T, Orawski AT, Simmons WH. Substrate specificity of aminopeptidase P from *Escherichia coli*: Comparison with membrane-bound forms from rat and bovine lung. Arch Biochem Biophys 1994; 311:28-34.

12. Goldberg AL. The mechanism and functions of ATP-dependent proteases in bacterial and animal cells. Eur J Biochem 1992; 203:9-23.

13. Ciechanover A. The ubiquitin-proteasome proteolytic pathway. Cell 1994; 79:13-21.

14. Yeager CL, Ashmun RA, Williams RK et al. Human aminopeptidase N is a receptor for human coronavirus 229E. Nature 1992; 357:420-2.

15. Delmas B, Gelfi JL, Haridon R et al. Aminopeptidase N is a major receptor for the enteropathogenic coronavirus TGEV. Nature 1992; 357:417-20.

16. Stirling CJ, Colloms SD, Collins JF et al. *xerB*, an *Escherichia coli* gene required for plasmid ColE1 site-specific recombination, is identical to *pepA*, encoding aminopeptidase A, a protein with substantial similarity to bovine lens leucine aminopeptidase. EMBO J 1989; 8:1623-7.

17. McCulloch R, Burke ME, Sherratt DJ. Peptidase activity of *Escherichia coli* aminopeptidase A is not required for its role in *cer* site-specific recombination. Mol Microbiol 1994; 12:241-51.

18. Delmas B, Gelfi J, Kut E et al. Determinants essential for the transmissible gastroenteritis virus-receptor interaction residue within a domain of aminopeptidase-N that is distinct from the enzymatic site. J Virol 1994; 68:5216-24.

19. Linderstrom-Lang K, Sato M. Die Spaltung von Glycylglycin, Alanylglycin und Leucylglycin durch Darm- und Malz- peptidasen. Hoppe-Seyler's Z Physiol Chem 1929; 104:83-90.

20. Mikola L, Mikola J. Occurrence and properties of different types of peptidase in higher plants. In: Dalling MJ, ed. Plant Proteolytic Enzymes. Vol I. Boca Raton, FL: CRR Press; 1986:97-117.

21. Preston KR, Kruger JE. Mobilization of monocot protein reserves during germination. In: Dalling MJ, ed. Plant Proteolytic Enzymes, Vol II. Boca Raton, FL: CRC Press; 1986:1-18.

22. Wilson KA. Role of proteolytic enzymes in the metabolism of protein reserves in the germinating dicot seed. In: Dalling MJ. ed. Plant Proteolytic Enzymes. Vol II. Boca Raton, FL: CRC Press, 1986:19-48.

23. Thayer SS, Choo HT, Rausser S et al. Characterization and subcellular localization of aminopeptidase in senescing barley leaves. Plant Physiol 1988; 87:894-7.

24. Sopanen T, Carfantan N. Activities of various peptidases in the senscesing petals of tulip. Physiol Plant 1976; 36:247-50.

25. Carrasco P, Carbonell J. Changes in the level of peptidase activities in pea ovaries during senescence and fruit set induced by gibberellic acid. Plant Physiol 1992; 92:1070-4.

26. Waters SP, Dalling MJ. Isolation and some properties of an aminopeptidase from the primary leaf of wheat (*Triticum aestivum* L.) Plant Physiol 1984; 75:118-24.

27. van der Valk HCPM, van Bentum MIA, van Loon LC. Proteolytic enzymes in developing leaves of oats (*Avena sativa* L.) II. Aminoacyl-2-naphthylamidases. J Plant Physiol 1989; 135:489-94.

28. Vodkin L, Scandalios JG. Developmental expression of genetically defined peptidases in maize. Plant Physiol 1979; 63:1198-204.

29. Capocchi L, Galleschi L, Meletti P. Proteolytic activities in the developing seeds of *X Haynaldoticum sardoum*. Biol Plant (PRAHA) 1990; 32:436-44.

30. Galleschi L, Schiano E, Izzo R et al. Changes in protease activities during the development of peach mesocarp. Plant Physiol Biochem 1991; 29:531-6.

31. Pearce G, Strydom D, Johnson S et al. A polypeptide from tomato leaves induces wound-inducible proteinase inhibitor proteins. Science 1991; 253:895-7.

32. McGurl B, Pearce G, Orozco-Cardenas M et al. Structure, expression, and antisense inhibition of the systemin precursor gene. Science 1992; 255:1570-3.

33. Schaller A, Ryan CA. Identification of a 50 KDa systemin-binding protein in tomato plasma membranes having Kex2-like properties. Proc Natl Acad Sci USA 1995; 91:11802-6.

34. Pautot V, Holzer FM, Reisch B et al. Leucine aminopeptidase: An inducible component of the defense response in *Lycopersicon esculentum* (tomato). Proc Natl Acad Sci USA 1993; 90:9906-10.

35. Hildmann T, Ebneth M, Pena-Cortes H et al. General roles of abscisic and jasmonic acids in gene activation as a result of mechanical wounding. Plant Cell 1992; 4:1157-70.

36. Rogers JC, Dean D, Keck GR. Aleurain: a barley thiol protease closely related to mammalian cathepsin H. Proc Natl Acad Sci USA 1985: 82:6512-6.

37. Bartling D, Weiler EW. Leucine aminopeptidase from *Arabidopsis thaliana*: Molecular evidence for a phylogenetically conserved enzyme of protein turnover in higher plants. Eur J Biochem 1992; 205:425-31.

38. Barrett AJ. Classification of peptidases. Meth Enzymol 1994; 244: 1-15.

39. Barrett AJ. An introduction to proteinases. In: Barrett AJ, Salvesen G, eds. Proteinase Inhibitors. Research Monographs in Cell and Tissue Physiology. Vol 12. Amsterdam: Elsevier, 1986:3-22.

40. Taylor A. Aminopeptidases: Structure and function. FASEB J 1993; 7:290-8.

41. Taylor A. Aminopeptidases: towards a mechanism of action. Trends Biochem Sci 1993; 205:168-74.

42. McDonald JK, Barrett AJ, eds. Mammalian Proteases. Volume 2. Exopeptidases. New York: Academic Press, 1986.

43. Matsushima M, Takahashi T, Ichinose M et al. Structural and immunological evidence for the identity of prolyl aminopeptidase with leucyl aminopeptidase. Biochem Biophys Res Commun 1991; 178:1459-64.

44. Sanderink G-J, Artur Y, Siest G. Human aminopeptidases: A review of the literature. J Clin Chem Biochem 1988; 26:795-807.

45. Wendel JF, Weeden NF. Visualization and interpretation of plant isozymes. In: Soltis DE, Soltis PS, eds. Isozymes in Plant Biology. Portland, Oregon: Dioscorides Press, 1989:5-45.

46. Kohno H, Kanda S, Kanno T. Immunoaffinity purification and characterization of leucine aminopeptidase from human liver. J Biol Chem 1986; 261:10744-8.

47. Burley SK, David RR, Taylor A et al. Molecular structure of leucine aminopeptidase at 2.7 Å resolution. Proc Natl Acad Sci USA 1990; 87:6878-82.

48. Wood DO, Solomon MJ, Speed RR. Characterization of the *Rickettsia prowazekii pepA* gene encoding leucine aminopeptidase. J Bacteriol 1993; 175:159-65.

49. Wynn EK, Murray DR. Aminopeptidases isolated from cotyledons of cowpea, *Vigna unguiculata*. Ann Bot 1985; 56:55-60.

50. Malik NSA, Pfeiffer NE, Williams DR et al. Peptidhydrolases of soybean root nodules. Identification, separation, and partial characterization of enzymes from bacteriod-free extracts. Plant Physiol 1981; 68:386-92.

51. Elleman TC. Aminopeptidases of pea. Biochem J 1974; 141:113-8.

52. Collier MD, Murray DR. Leucyl-β-naphthylamidase activities in developing seeds and seedlings of *Pisum sativum* L. Aust J Plant Physiol 1977; 4:571-82.

53. Tazaki K, Ishikura N. Purification and characterization of an aminopeptidase, LAPase 2, from *Euonymous* leaves. Plant Cell Physiol 1984; 25:731-7.

54. Couton JM, Sarath G, Wagner FW. Purification and characterization of a soybean cotyledon aminopeptidase. Plant Sci 1991; 75:9-17.

55. Kolehmainen L, Mikola J. Partial purification and enzymatic properties of an aminopeptidase from barley. Arch Biochem Biophys 1971; 145:632-42.

56. Ninomiya K, Tanaka S, Kawata S et al. Specificity of *Prunus* aminopeptidase toward peptide substrates. Agric Biol Chem 1981; 45:2121-2.

57. Ninomiya K, Tanaka S, Kawata S et al. Purification and properties of an aminopeptidase from seeds of japanese apricot. J Biochem 1981; 89:193-201.

58. Yamaoka Y, Takeuchi M, Morohashi Y. Purification and partial characterization of an aminopeptidase from mung bean cotyledons. Physiologia Plantarum 1994; 90:729-33.

59. Vodkin L, Scandalios JG. Comparative properties of genetically defined peptidases in maize. Biochem 1980; 19:4660-7.

60. Moriyasu Y, Sakano K, Tazawa M. Vacuolar/extravacuolar distribution of aminopeptidases in giant alga *Chara australis* and partial purification of one such enzyme. Plant Physiol 1987; 84:720-5.

61. Moriyasu Y, Miyoshi Y. Partial purification and characterization of aminopeptidase II from *Chara australis*. Plant Physiol 1989; 89:687-91.

62. Ashton FM, Dahmen WJ. A partial purification and characterization of two amino peptidases from *Cucurbita maxima* cotyledons. Phytochem 1967; 6:641-53.

63. Lin Y-H, Chan H-Y. An aminopeptidase (AP1) from sprouts of sweet potato (*Ipomoea batatas* (L.) Lam. cv. Tainong 64). Bot Bull Acad Sin 1992; 33:253-61.

64. Lin Y-H, Chan H-Y. An aminopeptidase (AP2) from sprouts of sweet potato (*Ipomoea batatas* (L.) Lam. cv. Tainong 64). Bot Bull Acad Sin 1992; 33:263-9.

65. Doi E, Shibata D, Matoba T et al. Activation of rice aminopeptidase by anions and some properties of the enzyme. Agric Biol Chem 1980; 44: 923-4.

66. du Toit PJ, Schabort JC. An aminopeptidase from *Agave americana*, chemical properties of the enzyme. Phytochem 1978; 17:371-6.

67. du Toit PJ, Schabort JC, Kempf PG et al. An aminopeptidase from *Agave americana*. Isolation and physical characterization. Phytochem 1978; 17:365-9.

68. Senkpiel K, Richter I, Beitz A et al. Aminopeptidase pattern of

Euglena gracilis. Biochem Physiol Pflanzen 1975; 168:585-95.

69. Pallavicini C, Peruffo ADB, Santoro M. Isolation and partial characterization of grape aminopeptidase. J Agric Food Chem 1981; 1216-20.

70. Johnson R, Storey R. Aminopeptidase activity from germinated jojoba cotyledons. Plant Physiol 1985; 79:641-5.

71. Koebner RMD, Martin PK. Chromosomal control of the aminopeptidase of wheat and its close relatives. Theor Appl Genet 1989; 78:657-64.

72. Mikola J. Activities of various peptidases in cotyledons of germinating peanut (*Archis hypogaea*). Physiol Plant 1976; 36:255-8.

73. Salmia M, Mikola J. Activities of two peptidases in resting and germinating seeds of Scots pine, *Pinus sylvestris.* Physiol Plant 1975; 33:261-5.

74. Caldwell JB, Sparrow LG. Purification and characterization of an unusual aminopeptidase from pea seeds. Aust J Plant Physiol 1980; 7:131-40.

75. Liu X-Q, Jagendorf AT. Neutral peptidases in the stroma of pea chloroplasts. Plant Physiol 1986; 81:603-8.

76. Blattler R, Feller U. Identification and stability of aminopeptidases in extracts of bean seeds. Aust J Plant Physiol 1988; 15:613-9.

77. Kermode AR, Gifford DJ, Thakore E et al. On the composition, deposition and mobilization of proteins in the cotyledon of castor bean (*Ricinus communis* L. cv. Hale) seeds: Their role as storage proteins. J Exp Bot 1985; 36:792-9.

78. Ninomiya K, Tanaka S, Kawata S et al. Substrate specificity of a proline iminopeptidase from apricot seeds. Agric Biol Chem 1983; 47:629-30.

79. Waters SP, Dalling MJ. Purification and characterization of an iminopeptidase from the primary leaf of wheat (*Triticum aestivum* L.) Plant Physiol 1983; 73:1048-54.

80. Ninomiya K, Kawatani K, Tanaka S et al. Purification and properties of a proline aminopeptidase from apricot seeds. J Biochem 1982; 92:413-21.

81. Senkpiel VK, Richter I, Barth A. Description of a proline aminopeptidase from *Euglena gracilis.* Biochem Physiol Pflanzen 1974; 166:7-21.

82. Lang WC, Blatt D, Plapp R. Proteolytic enzymes in *Chlamydomonas.* I. A survey on the aminopeptidase pattern in asynchronous vegetative cells of *Chlamydomonas reinhardii.* Plant Cell Physiol 1979; 20:657-65.

83. Kieliszewski MJ, Lamport DTA. Extensin: repetitive motifs, functional sites, post-translational codes, and phylogeny. Plant J 1994; 5:157-72.

84. Altan D, Gilbert C, Blanc B et al. Cloning, sequencing and char-

acterization of the *pepIP* gene encoding a proline iminopeptidase from *Lactobacillus delbrueckii* subsp. *bulgaricus* CNRZ 397. Microbiol 1994; 140:527-35.

85. Kitazano A, Yoshimto T, Tsuru D. Cloning, sequencing, and high expression of the proline iminopeptidase gene from *Bacillus coagulans*. J Bacteriol 1992; 174:7919-25.

86. Simmons WH, Orawski AT. Membrane-bound aminopeptidase P from bovine lung. Its purification, properties, and degradation of bradykinin. J Biol Chem 1992; 267:4897-903.

87. Mikkonen A, Mikola J. Separation and partial characterization of two alkaline peptidases from cotyledons of resting kidney beans, *Phaseolus vulgaris*. Physiol Plant 1986; 68:81-85.

88. Mikkonen A. Purification and characterization of leucine aminopeptidase from kidney bean cotyledons. Physiol Plant 1992; 84:393-8.

89. El Amrani A, Couée I, Carde J-P et al. Modification of etioplasts in cotyledons during prolonged dark growth of sugar beet seedlings. Identification of etiolation-related plastidial aminopeptidase activities. Plant Physiol 1994; 106:1555-65.

90. Berger J, Johnson MJ. The leucylpeptidases of malt, cabbage, and spinach. J Biol Chem 1939; 130:655-7.

91. Ashton FM, Dahmen WJ. Purification and characterization of an aminopeptidase from *Cucurbita maxima* cotyledons. Phytochem 1968; 7:189-97.

92. Casano LM, Desimone M, Trippi VS. Proteolytic activity at alkaline pH oat leaves, isolation of an aminopeptidase. Plant Physiol 1989; 91:1414-8.

93. Sopanen T, Lauriere C. Activities of various peptidases in the first leaf of wheat. Physiol Plant 1976; 36:251-4.

94. Sopanen T. Purification and partial characterization of a dipeptidase from barley. Plant Physiol 1976; 57:867-71.

95. Sopanen T, Mikola J. Purification and partial characterization of barley leucine aminopeptidase. Plant Physiol 1975; 55:809-14.

96. Ashton FM, Dahmen WJ. Purification and characterization of a dipeptidase from *Cucurbita maxima* cotyledons. Phytochem 1967; 6:215-25.

97. Galleshi L, Pellegrini L. Purification and some properties of an aminopeptidase from endosperms of X *Haynaldoticum sardoum* Meletti et Onnis. New Phytol 1989; 113:301-6.

98. Taylor A, Peltier CZ, Torre FJ et al. Inhibition of bovine lens leucine aminopeptidase by bestatin: Number of binding sites and slow binding of this inhibitor. Biochem 1993; 32:784-90.

99. Kim H, Lipscomb WM. Structure and mechanism of bovine lens leucine aminopeptidase. Adv Enzymol Rel Areas Molec Biol 1994; 68:153-213.

100. Melbye SW, Carpenter FH. Leucine aminopeptidase (bovine lens).

J Biol Chem 1976; 246:2459-63.

101. Ledeme N, Vincent-Fiquet O, Hennon G et al. Human liver L-leucine aminopeptidase: Evidence for two forms compared to pig liver enzyme. Biochimie 1983; 65:397-404.

102. Carpenter FH, Vahl JM. Leucine aminopeptidase (bovine lens). J Biol Chem 1973; 248:294-304.

103. Burley SK, David PR, Lipscomb WN. Leucine aminopeptidase: Bestatin inhibition and a model for enzyme-catalyzed peptidase hydrolysis. Proc Natl Acad Sci USA 1991; 88:6916-20.

104. Kim H, Lipscomb WN. X-ray crystallographic determination of the structure of bovine lens leucine aminopeptidase complexed with amastatin: Formulation of a catalytic mechanism featuring a *gem*-diolate transition state. Biochem 1993; 32:8465-78.

105. Kim H, Lipscomb WN. Differentiation and identification of the two catalytic metal binding sites in bovine lens leucine aminopeptidase by x-ray crystallography. Proc Natl Acad Sci USA 1993; 90:5006-10.

106. Cuypers HT, van Loon-Klaassen LAH, Vree Egberts WT et al. The primary structure of leucine aminopeptidase from bovine lens. J Biol Chem 1982; 257:7077-85.

107. Wallner BP, Hession C, Tizard R et al. Isolation of bovine kidney leucine aminopeptidase cDNA: Comparison with the lens enzyme and tissue-specific expression of two mRNAs. Biochemistry 1993; 32:9296-301.

108. Bartling D, Nosek J. Molecular and immunological characterization of leucine aminopeptidase in *Arabidopsis thaliana*: A new antibody suggests a semi-constitutive regulation of a phylogenetically old enzyme. Plant Sci 1994; 9:199-209.

109. Bray E. Molecular responses to water deficit. Plant Physiol 1993; 103:1035-40.

110. Peña-Cortés H, Willmitzer L, Sanchez-Serrano JJ. Abscisic acid mediates wound induction but not developmental-specific expression of the proteinase inhibitor II gene family. Plant Cell 1991; 3:963-972.

111. Herbers K, Prat S, Willmitzer L. Functional analysis of leucine aminopeptidase from *Solanum tuberosum* L. Planta 1994; 194:230-40.

112. Harris CA, Hunte B, Krauses MR et al. Induction of leucine aminopeptidase by interferon-γ Identification by protein microsequencing after purification by preparative two-dimensional gel electrophoresis. J Biol Chem 1992; 267:6865-9.

112a.Gu Y-Q, Pautot V, Holzer FM et al. A complex array of proteins related to the multimeric leucine aminopeptidase of tomato. Plant Physiol 110:1257-1266.

113. Holwerda BC, Galvin NJ, Baranski TJ et al. In vitro processing of aleurain, a barley vacuolar thiol protease. Plant Cell 1990; 2:1091-106.

114. Holwerda BC, Rogers JC. Purification and characterization of

aleurain. A plant thiol protease functionally homologous to mammalian cathepsin H. Plant Physiol 1992; 99:848-55.

115. Schwartz WN, Barrett AJ. Human cathepsin H. Biochem J 1980; 191:487-97.

116. Rothe M, Zichner A, Auerswald EA et al. Structure/function implications for the aminopeptidase specificity of aleurain. Eur J Biochem 1994; 224:559-65.

117. Watanabe H, Abe K, Emori Y et al. Molecular cloning and gibberellin-induced expression of multiple cysteine proteinases of rice seeds (oryzains). J Biol Chem 1991; 266:16897-902.

118. Sopanen T, Burston, D, Taylor E et al. Uptake of glycylgylcine by the scutellum of germinating barley grain. Plant Physiol 1978; 61:630-3.

119. Higgins CF, Payne JW. Peptide transport by germinating barley embryos: Uptake of physiological di- and oligopeptides. Planta 1978; 138:211-5.

120. Gifford DJ, Wenzel KA, Lammer DL. Lodgepole pine seed germination. I. Changes in peptidase activity in the megagametophyte and embryonic axis. Can J Bot 1989; 67:2539-43.

121. Kruger JE, Preston KR. Changes in aminopeptidases of wheat kernels during growth and maturation. Cereal Chem 1978; 55:360-72.

122. Leung DW, Bewley JD. Increased activity of aminopeptidases in the cotyledons of red light-promoted lettuce seeds is controlled by the axis. Physiol Plant 1983; 59:127-133.

123. Samac D, Storey R. Proteolytic and trypsin inhibitor activity in germinating jojoba seeds (*Simmondsia chinesis*). Plant Physiol 1981; 68:1339-44.

124. Cohen HP, deMaggio AE. Biochemistry of fern spore germination: Protease activity in ostrich fern spores. Plant Physiol 1986; 80:992-6.

125. Salmia MA, Mikola JJ. Localization and activity of naphthylamidases in germinating seeds of Scots pine, *Pinus sylvestris*. Physiol Plant 1976; 38:73-7.

126. Chrispeels MJ, Boulter D. Control of storage protein metabolism in the cotyledons of germinating mung beans: Role of endopeptidases. Plant Physiol 1975; 55:1031-7.

127. Crump JA, Murray DR. Proteolysis in cotyledon cells of *Phaseolus vulgaris* L.: Changes in multiple hydrolase activities following germination. Aust J Plant Physiol 1979; 6:467-74.

128. Feller U. Nitrogen mobilization and proteolytic activities in germinating and maturing bush beans (*Phaseolus vulgaris*). Z Pflanzenphysiol 1990; 95:413-22.

129. Waters SP, Dalling MJ. Distribution and characteristics of amino acyl-β-naphthylamidase activities in wheat seedlings. Aust J Plant Physiol 1979; 6:595-606.

130. Mikola J, Kolehmainen L. Localization and activity of various pep-

tidases in germinating barley. Planta 1972; 164:167-77.

131. Feller U, Soong T-ST, Hageman RH. Patterns of proteolytic enzyme activities in different tissues of germinating corn (*Zea mays* L.) Planta 1978; 140:155-62.

132. Hammerton RW, Ho T-HD. Hormonal regulation of the development of protease and carboxypeptidase activities in barley aleurone layers. Plant Physiol 1985; 80:692-7.

133. Pino E, Martin L, Guerra H et al. Effect of dihydrozeatin on the mobilization of protein reserves in proteins bodies during the germination of chick-pea seeds. J Plant Physiol 1991; 137:425-32.

134. Mitsuhashi W, Koshiba T, Minamikawa T. Influence of axis removal on amino-, carboxy- and endopeptidase activities in cotyledons of germinating *Vigna mungo* seeds. Plant Cell Physiol 1984; 25:547-54.

135. Murray DR. Nutritive role of the seedcoats during embryo development in *Pisum sativum* L. Plant Physiol 1979; 64:763-9.

136. Grierson D. Molecular biology of fruit ripening. Oxford Surv Plant Molec Cell Biol 1986; 3:363-83.

137. Galleshi L, Schiano E, Izzo R et al. Changes in protease activities during the development of peach mesocarp. Plant Physiol 1991; 29:531-6.

138. Murray DR, Peoples MB, Waters SP. Proteolysis in the axis of the germinating pea seed. I. Changes in protein degrading enzyme activities of the radicle and primary root. Planta 1979; 147:111-6.

139. Mullet JE. Chloroplast development and gene expression. Ann Rev Plant Physiol Plant Mol Biol 1988; 39:475-502.

140. Waters SP, Noble ER, Dalling MJ. Intracellular localization of peptide hydrolases in wheat (*Triticum aestivum* L.) leaves. Plant Physiol 1982; 69:575-9.

141. Keegstra K, Olsen LJ, Theg SM. Chloroplastic precursors and their transport across the envelope membranes. Ann Rev Plant Physiol Plant Molec Biol 1989; 40:471-501.

141a.Gu Y-Q, Chao WS, Walling LL. Localization and post-translational processing of the wound-induced leucine aminopeptidase proteins of tomato. J Biol Chem 1996; in press.

142. Michel H, Hunt DF, Shabanowitz J et al. Tandem mass spectroscopy reveals that three photosystem II proteins of spinach chloroplasts contain N-acetyl-O-phosphothreonine at their NH2 termini. J Biol Chem 1988; 263:1123-30.

143. Howe CJ, Fearnley IM, Walker JE et al. Nucleotide sequence of the genes for the alpha, beta, and epsilon subunits of wheat chloroplast ATP synthase. Plant Molec Biol 1985; 4:333-45.

144. Widger WR, Cramer WA, Hermodson M et al. Purification and partial amino acid sequence of the chloroplast cytochrome b-559. J Biol Chem 1985; 259:3870-6.

145. Farchaus J, Dilley W. Purification and partial sequence of the Mr

10,000 phosphoprotein from spinach thylakoids. Arch Biochim Biophys 1986; 244:94-101.

146. James F, Brouquisse R, Pradet A et al. Changes in proteolytic activities in glucose-starved maize root tips. Regulation by sugars. Plant Physiol Biochem 1993; 31:825-56.

147. Thomas H. Foliar senescence mutants and other genetic variants. In: Thomas H, Greirson S, eds. Developmental Mutants in Higher Plants. Cambridge, MA:Cambridge University Press, 1987:245-65.

148. Huffaker RC. Proteolytic activity during senescence of plants. New Phytol 1990; 116:199-231.

149. Buchanan-Wollaston V. Isolation of cDNA clones for genes that are expressed during leaf senescence in *Brassica napus*. Identification of a gene encoding a senescence-specific metallothionein-like protein. Plant Physiol 1994; 105:839-46.

150. Lamb CJ, Lawton MA, Dron M et al. Signals and transduction mechanisms for activation of plant defenses against microbial attack. Cell 1989; 56:215-224.

151. Bowles D. Defense-related proteins in higher plants. Ann Rev Biochem 1990; 59:873-907.

152. Vera P, Conejero V. The induction and accumulation of the pathogenesis-related P69 proteinase in tomato during citrus exocortis viroid infection and in response to chemical treatments. Physiol Molec Plant Path 1989; 34:323-34.

153. Luthe DS, Jiang B, Siregar U et al. Relationship of a 33 kDa putative cysteine proteinase with fall armyworm (*Spodoptera frugiperda*) resistance in maize. J Cell Biochem 1995; Supplement 19A:157.

154. Linthorst HJM, van der Does C, Brederode FT et al. Circadian expression and induction by wounding of tobacco genes for cysteine proteinase. Plant Molec Biol 1993; 21:685-94.

155. Trena L, Matousek J. Aminopeptidase activity in potato spindle viroid infected tomato leaves. Arch Phytopath Pflanzen 1991; 27:117-25.

155a. Schaller A, Bergey DR, Ryan CA. Induction of wound-response genes in tomato leaves by bestatin, an inhibitor of aminopeptidases. Plant Cell 1995; 7:1893-8.

156. Lightner J, Pearce G, Ryan CA et al. Isolation of signaling mutant of tomato (*Lycopersicon esculentum*). Mol Gen Genet 1993; 241:595-601.

157. Ryals J, Ward E, Ahl-Goy P et al. Systemic acquired resistance: an inducible defense mechanism in plants In: Wray JL, ed. Inducible Plant Proteins. Great Britain:University Press Cambridge, 1992: 205-229.

158. Templeton MD, Mitchell RE, Sullivan PA et al. The inactivation of ornithine transscarbamylase by N*-(N'-sulpho-diaminophosphinyl)-L-ornithine. Biochem J 198-; 228:347-52.

159. Mitchell RE, Bieleski RL. Involvement of phaseolotoxin in halo

blight of beans. Transport and conversion to functional toxin. Plant Physiol 1977; 60:723-729.

160. Kwok OCH, Patil SS. Activation of a chlorosis-inducing toxin of *Pseudomonas syringae* pv. *phaseolicola* by leucine aminopeptidase. FEMS Microbiol Letters 1982; 14:247-9.

161. van den Ackerveken GFJM, Vossen P, de Wit PJGM. The AVR9 race-specific elicitor of *Cladosporium fulvum* is processed by endogenous and plant proteases. Plant Physiol 1993; 103:91-6.

162. Mehdy MC. Active oxygen species in plant defense against pathogens. Plant Physiol 1994; 105:467-72.

163. Dean RT, Gieseg S, Davies MJ. Reactive species and their accumulation on radicle-damaged proteins. Trends Biochem Sci 18: 437-41.

164. Eckey-Kaltenbach H, Groskopf E, Kiefer E et al. Defense-related genes of parsley (*Petroselium crispum* L.) are induced by ozone. 4th International Inter Congress Plant Molec Biol 1994; Abstract 452.

165. Prasad TK, Anderson MD, Martin BA et al. Evidence for chilling-induced oxidative stress in maize seedlings and a regulatory role for hydrogen peroxidase. Plant Cell 1994; 6:65-74.

166. Dubey RS, Rani M. Influence of NaCl salinity on the behavior of protease, aminopeptidase and carboxypeptidase in rice seedlings in relation to salt tolerance. Aust J Plant Physiol 1990; 17:215-21.

167. Dubey RS, Rani M. Influence of sodium choride salinity on peptidase activities and the status of total amino acids in germinating rice seeds of differing salt tolerance. Trop Sci 1990; 30:133-45.

INDEX

A

AAP1, 108, 111-112, 121
acetylation, 92
actin, 92
Aeromonas, 4, 11, 36, 48, 53, 56, 72
alanyl AP, 70-72
aleurain, 175, 190-191, 194, 207
amastatin, 3-4, 28, 35-37, 43-44, 47-48, 52-53, 59, 79, 80, 144, 157, 160, 166- 167, 186
aminopeptidase(s)
    function, 4, 7, 10, 21-22, 60, 87, 91
    physiology, 60
    plant, 191, 205-206
    structure, 4, 21
aminopeptidase A, 7-8
aminopeptidase I, 112, 112-114
aminopeptidase M, 91, 136-137, 139-140
aminopeptidase N, 7-8, 10, 109-111, 136, 174, 201
aminopeptidase P, 100
aminopeptidase Y, 114
aminopeptidase yscII, 71, 109-112, 121
amphiphilic α-helix, 112
antitumor antibiotic, 108
APE2, 4, 108-109
apoenzyme, 77, 82-83, 85, 137
arginine aminopeptidase, 111,143, 146
azidobestatin, 31, 35, 37

B

B. subtilis, 56, 74-75, 85, 101, 116
bestatin, 3, 6-7, 11, 21, 23, 28-37, 41, 43-44, 49, 51, 53, 56, 59, 79-80, 114
bleomycin hydrolase, 6, 56, 108-109, 122
BLH1, 6, 108-109, 122
bovine lens AP (blAP), 21-60, 72-86

C

catalytic mechanism, 76, 84, 140, 186
Co2+, 4, 45, 48, 114-115
cobalt, 36, 45, 74, 78-79, 82, 85-86, 101, 117
cocatalytic zinc site, 72
coordination sphere, 79, 84-85
creatinase, 100
CUP1, 111
cysteine proteinase, 56, 108

D

deformylase enzyme, 94
degradation, 1-2, 6-9, 91-92, 103, 112, 157, 173-174, 192-193, 199
dependent metallopeptidase, 110
dipeptidase, 75, 131, 173, 180, 184, 193

E

E. coli, 5-6, 10, 22, 48, 52-54, 56, 70-75, 85, 93, 95, 98, 101, 110, 116, 119, 188-189
    aminopeptidase N, 111
    MetAP, 95-97, 115, 120
eIF2a eukaryotic initiation, 100
epitope-tagging/immunoprecipitation approach, 120
erythrocyte, 133, 138, 147

F

fruit maturation, 175, 194

G

glycogen accumulation, 111, 123
glycopeptide antibiotics, 109

H

hemagglutinin epitope, 120
human cathepsin H, 175, 190
human intestine aminopeptidase N, 111
human leukotriene A4 hydrolase, 71, 110, 111

I

initiator methionine, 93, 103, 118, 121

L

leucine aminopeptidase, 72, 91, 109, 174-176, 181, 184-185, 187-188, 193, 202, 207
    lens (LAP), 21-22
    plant, 185-190
leucinephosphonic acid, 28, 35
leukotriene A_4, 129-149
    hydrolase, 7, 56, 70-71, 74, 110, 111, 136-149

M

M1 family, 70-77
M17 family, 72-73
mechanism of action, 49-51, 59, 80
metal ion(s), 2, 4, 45-49, 79, 82-83, 117
    content, 2, 4, 34
metalloenzymes, 69, 85, 92, 157, 182, 184
metallohybrid, 80, 82-83
metalloprotease, 121
MetAP families, 101-102
MetAP1, 115-119, 121-122
MetAP2, 119-122
MetAPs, 91-104, 115-116, 118-122
methionine aminopeptidase(s), 3, 5-8, 48-49, 91-107,
    115, 120-123, 160, 173-174
microsomal fraction, 92
mitochondria, 3, 5, 92-93, 112, 197
myristoylation, 5, 118

N

N-end rule, 9, 103
N-formyl methionine, 92
NH2-terminal methionine, 5, 115, 118
nonspecific aminopeptidase, 107-112, 121

O

O-linked GlcNAc, 120

P

physiological function, 1, 5-8, 102, 121
polylysine blocks, 121
polyubiquitination, 103
porcine kidney AP, 72
pre-B cell antigen (BP1), 110-111
prolidase, 100
prolyl AP, 5, 51
proteasome, 4, 107-109
protein synthesis, 5, 10, 58, 91-92, 100, 103, 107,
    120, 194, 197
protein/nucleic acid interactions, 121
protein/protein interactions, 121
PUT2, 111

R

rapidly exchanging site, 77, 83
rat p67, 100, 119, 120, 121, 122
RING finger, 116

S

Saccharomyces cerevisiae, 6-7, 85, 107, 112, 114-115,
    118, 174, 187
Salmonella typhimurium, 5, 53, 174
seed germination, 174, 191-192
seedling development, 175, 195, 197
senescence, 174, 187, 196, 199, 200, 201
sequence homology, 52, 53, 56, 73, 98
site-directed mutagenesis, 74, 93, 117, 137, 147
slowly exchanging site, 78, 80, 83
Streptococcus thermophilus CNRZ302, 56
substrate analog, 2, 36, 49
Sui3, 121
suicide, 147
suppressor gene, 120

T

tetrad analysis, 121
thermolysin, 56, 70-71, 74, 76, 135-137
translational regulation, 123
two-cobalt cluster, 74

U

ubiquitin ligase, 103

V

vacuolar aminopeptidase, 107, 112-115

X

X-ray structure , 60

Y

yeast
    aminopeptidase yscII, 71, 111-112
    MetAp, 93-97
    vacuolar aminopeptidase I, 112-113

Z

zinc binding domains, 111
zinc dependent metallopeptidase, 110
zinc fingers, 56, 116-118
Zn^{2+} fingers, 7